I0485785

Problem Solving in Engineering Hydrology

By

Dr. Eng. Faris Gorashi Faris

Prof. Dr. Eng. Isam Mohammed Abdel-Magid Ahmed

Dr. Mohammed Isam Mohammed Abdel-Magid

Second edition enhanced, improved and updated, 2015.
Second revised edition, 2013 (Electronically).

First edition printed by: University of Dammam Press, Dammam, 2013.

ISBN-13: 978-1516876945
ISBN-10: 1516876946

Dr. Faris Gorashi Faris: Deputy Dean, Faculty of Engineering and Technology Infrastructure, (Infrastructure University Kuala Lumpur, (IUKL), Jalan Ikram – Uniten, 43000 Kajang , Selangor MALAYSIA, Fax : +603-89256361, Tel: +603-8738 3388 Ext 312 (off), +6019 640 7203, E.mail: dr.faris@gmail.com, faris@iukl.edu.my

Prof. Dr. Eng. Isam Mohammed Abdel-Magid Ahmed: Head Proofreading and revision department at the Centre of Scientific Publications and Dammam University, Professor of water resources and environmental engineering, Building 800, Room 240 Environmental Engineering Department, College of Engineering, University of Dammam, Box 1982, Dammam 31451, KSA, Fax: 96638584331+, Phone: 966530310018+, E-mail: iahmed@ud.edu.sa, isam_abdelmagid@yahoo.com, Web site: http://www/sites.google.com/site/isamabdelmagid/

Dr. Mohammed Isam Mohammed Abdel-Magid: Department of Internal Medicine, Khasab Hospital & Polyclinic, P. O. Box 306, Postal 811, Khasab, Musandam, Sultanate of OmanPhone: +97470445235, +971558215655, +249969263307, E-mail: mohammed_isam1984@yahoo.com, website: http://sites.google.com/site/mohammedisam2000

Preface for Second Edition

The first edition of "Problem solving in engineering hydrology" was printed by Dammam University Press. An updated second version was uploaded electronically for wider spreading and circulation. The first edition of the book was used as a supporting document and a course supplement to the subject of engineering hydrology launched to students at the Department of Environmental Engineering, College of Engineering of Dammam University. Extensive use of the book by concerned students throughout the course lectures and tutorials delivery proved its fruitful value and appeal. Students praised its added academic value, alleviated standards; achieving course set objectives and attaining significant benefits.

Positive comments, constructive remarks, sound contributions and valid criticisms were received from other users of the book at other higher education institutions, engineering associations and firms locally, regionally and at IUKL Malaysia. This recognition and reputation encouraged the authors to seek wider book distribution and better use by other colleagues, undergraduate students at engineering colleges and schools, postgraduate scholars, civil engineers, hydrologists, personnel working at sectors concerned with engineering hydrology, research institutions, education enterpises and other related fields and disciplines.

The new edition of the book hosted the incorporation of computer model programs for the different hydrological scenarios and encountered problems presented throughout the book. Developed programs were coded with Microsoft Visual Basic.NET 10 programming language, using Microsoft Visual Studio 2010 Professional Edition. Most of the examples herein have an equivalent code listed alongside through the text. To avoid repetition though, some example programs were omitted whenever there was resemblance to another example elsewhere, to which the reader is kindly requested to refer to.

It is hoped that this new edition of the manuscript would be a major revision of the popular "problem solving in engineering hydrology" book.

The authors are very grateful to all those who helped in many ways, through advice in person or though their published scientific work, students with whom they enjoyed studying engineering hydrology and for those who offered fruitful suggestions, prolific inputs and splendid contributions.

Dr. Eng. Faris Gorashi Faris
Prof. Dr. Eng. Isam Mohammed Abdel-Magid Ahmed Sr.
Dr. Mohammed Isam Mohammed Abdel-Magid Jr.
KSA, Malaysia, Sultanate of Oman, 2015

Preface for First Edition

Engineering hydrology books, manuscripts and text material are abundant addressing perceptions, fundamentals, methods, models, designs and associated scientific concepts. However, rare are the books that deal with problem solving and practical complications in the field of engineering hydrology.

This book has been written to tackle material described in the 2-credit hour course of "Engineering Hydrology, ENVEN 431" given during the first semester for Junior Engineering students of the department of environmental engineering of the College Of Engineering at Dammam University as well as a 3-credit course Hydraulics and Hydrology, BEC 208 offered in the second year for Bachelor of Civil Engineering students at the Infrastructure University Kuala Lumpur. The named courses contents encompassed: Hydrologic cycle, precipitation and water losses. Catchment's characteristics and runoff processes. Flood estimation and control. Hydrograph analysis. Flood & reservoir routing. Groundwater occurrence, distribution, movement, exploration and recharge, well hydraulics and design, interaction of ground and surface water. Differential equations of groundwater flow. Darcy Law, solutions of the steady and unsteady flow, differential equations for confined and unconfined flows. Pumping test design. Groundwater models, leaky aquifers. Saltwater intrusion.

The book also covered the course contents of the subject "Hydrology and water gauging" CIVL 561 offered for the fifth year students of the department of civil engineering of Sudan University for Science and Technology. This particular course covered: Elements of hydrology. Precipitation. Surface runoff. Evaporation and evapotranspiration. Infiltration and groundwater flow. Hydrological measurement networks. Hydrograph. Flood routing.

Likewise, the book covered "hydrology" Course contents offered at the civil engineering department of the college of engineering at United Arab Emirates University.

The book also covered material subject dealt with while teaching the course in "hydrology" at School of Engineering and Technology Infrastructure, Kuala Lumpur Infrastructure University College.

Objectives of the book are meant to fulfill the main learning outcomes for students registered in named courses, which covered the following:

a) Solving problems in hydrology and making decisions about hydrologic issues that involve uncertainty in data, scant/incomplete data, and the variability of natural materials.
b) Designing a field experiment to address a hydrologic question.
c) Evaluating data collection practices in terms of ethics.
d) Interpret basic hydrological processes such as groundwater flow, water quality issues, water balance and budget at a specific site at local and regional scales based on available geological maps and data sets.
e) Conceptualizing hydrogeology of a particular area in three dimensions and be able to predict the effects on a system when changes are imposed on it.

Learning outcomes are expected to include the following:

a) Overview of essential concepts encountered in hydrological systems.
b) Developing a sound understanding of concepts as well as a strong foundation for their application to real-world, in-the-field problem solving.
c) Acquisition of knowledge by learning new concepts, and properties and characteristics of water.
d) Cognitive skills through thinking, problem solving and use of experimental work and inferences
e) Numerical skills through application of knowledge in basic mathematics and supply issues.
f) Student becomes responsible for their own learning through solution of assignments, laboratory exercises and report writing.

Knowledge to be acquired from the course is expected to incorporate:

a) Use real world data to develop a water budget for unfamiliar basins.
b) Interpret groundwater resources data and correlate the geology with the groundwater regime.
c) Evaluate ways in which water influences various geologic processes
d) Identify interconnections in hydrological systems and predict changes.
e) Predications of hydrological terms influencing the hydrological cycle.
f) Collection of data, analysis and interpretation

Cognitive skills to be developed are expected to incorporate the following:
a) Capturing ability of reasonable scientific judgment and concepts of appropriate decision making.
b) Students will be able to apply the knowledge of hydrology that they have learnt in this course in practical environmental engineering domain.
c) Students should be able to design and apply necessary procedures and precautions to produce durable hydro logical systems.

The majority of incorporated problems within the book represent examination problems that were given by the author to students at University of United Arab Emirates (UAE), Sudan University for Science and Technology (SUST), University of Dammam (UoD) and Infrastructure University Kuala Lumpur (IUKL). This is to expose students to different and design constraints and prevailing settings.

"Problem solving in engineering hydrology" is primarily proposed as an addition and a supplementary guide to fundamentals of engineering hydrology. Nevertheless, it can be sourced as a standalone problem solving text in engineering hydrology. The book targets university students and candidates taking first degree courses in any relevant engineering field or related area. The document is valued to have esteemed benefits to postgraduate students and professional engineers and hydrologists. Likewise, it is expected that the book will stimulate problem solving learning and quicken self-teaching. By writing such a script it is hoped that the included worked examples and problems will guarantee that the booklet is a precious asset to student-centered learning. To achieve such objectives immense care was paid to offer solutions to selected problems in a well-defined, clear and discrete layout exercising step-by-step procedure and clarification of the related solution employing vital procedures, methods, approaches, equations, data, figures and calculations.

The authors acknowledge support, inspiration and encouragement from many students, colleagues, friends, institutions and publishers. The authors salute the motivation and stimulus help offered to them by Dean Dr. Abdul-Rahman ben Salih Hariri, Dean College of Engineering of University of Dammam.. Thanks are also extended to Miss Azlinda binti Saadon from the Faculty of Civil Engineering at The Infrastructure University Kuala Lumpur for her cooperation and support. Special and sincere vote of thanks would go to Mr. Mugbil ben Abdullah Al-Ruwais the director of Dammam University Press and his supreme staff for their patience, dedication and orderly typing of the book

Dr. Eng. Faris Gorashi Faris

Prof. Dr. Eng. Isam Mohammed Abdel-Magid Ahmed

Malaysia, KSA, 2013

Table of contents

List of Tables

List of Figures

List of Appendices

Abbreviations, notations, symbols and terminology used in the book

A = Area, Area of groundwater basin normal to direction of water flow, Drainage area, Area of catchment basin (m^2, km^2, ha).

A = Cross-sectional area
A_i = Area of polygon surrounding station number i, located in its middle, m^2.
a = An exponent varying with surface roughness and stability of the atmosphere.
a = Index of surface connected porosity.
a, b, c = Hydraulic monitoring (gauging) stations to measure rainfall.
a = Rain slope constant.
a = Artificial recharge or loss of irrigation.
a and b = Constants.
a = Height (m) between zero on the gauge and the elevation of zero flow.
B = Heat exchange between soil and surface.
B = Stream roughness factor.
b = An empirical constant.
C = Coefficient representing the ratio of runoff to rainfall.
C = An empirical constant, Constant.
C = Coefficient controlling rate of decrease of loss-rate function.
C = Rational coefficient of surface flow.
C = Weight of salt solution passing sampling point per second.
C = Chezy roughness coefficient
C_2 = Weight of salt passing sampling point per second.

C_b = Concentration of tracer element in the river at start of injection.

C_i = Concentration of injected tracer element (within the stream).

12

C_m = Concentration of tracer component at measurement point (at equilibrium).

C_p = Specific heat of air at constant pressure.

C_t = Coefficient depending on units of drainage basin characteristics.
C_1 = Weight of concentrated solution added per second.
c, n = Locality constants.
c_p = Constant.

D = Distance from ocean, m.

D = Discharge from stage, m.

D = Rate on outflow, m^3/s.

D_x = Time period selected, m/s.

d = Surface-layer depth.

dh/dt = Change in stage during measurement, m/s.

dP = Pressure change with temperature, Pa.

dS/dt = Time rate of change of storage.

E = Amount of evaporation.

E = Evaporation from surface of river basin.

E = Evapotranspiration.

E = Amount of evaporation during a month, mm, inches.

E = Actual vapor pressure, mmHg.

E = Exponent, constant.

E = Actual vapor pressure of air at temp t.

EV = Total evaporation from soil and plants for specified time period.

EV = Evaporation, cm.

E_a = Evaporation from the earth, and from storage in low-lying areas.

E_a = Open water evaporation per unit time, mm/day.

E_a = Evaporation, mm/day.

E_a = Aerodynamic term.

E_b = Net energy lost by body of water through exchange of long wave radiation between atmosphere and body of water.

E_e = Energy utilized for evaporation.

E_h = Energy conducted from body of water to atmosphere as sensible heat.

E_o = Evaporation, mm/day.

E_o = Evaporation of lake, mm/day.

E_q = Increase in stored energy in the body of water.

E_r = Reflected solar radiation.

E_s = Solar radiation incident to water surface.

E_s = Saturated vapor pressure.

E_T = Evaporation from open surface of water (or equivalent in heat energy).

E_T = Evaporation rate.

E_v = Net energy adverted into the body of water.

E_w = Saturation vapor pressure at temp t_w of surface water of lake, mmHg.

e = Median value for vapor pressure, mbar.

e = Vapor pressure.

e = Base of natural logarithms.

e = Actual vapor pressure.

e = Actual vapor pressure at a defined height above surface.

e = Vapor pressure of air (monthly average), inches Hg.

e = Natural algorithm base.

e_s = Median value for saturation pressure, mbar.

e_a = Vapor pressure of air, mbar.

e_s = Saturation vapor pressure.

e_s = Saturated vapor pressure (mbar) when temperature is T_w

e_s = Saturated vapor pressure at surface temperature.

e_s = Saturation flexibility at surface, mm Hg.

e_s - e = d, = Lack of saturation Millbar.

e_w = Partial pressure of the gas for wet bulb temperature.

e_s = Saturation vapor pressure of air when temperature (t) ° C, mbar, mm Hg.

e_s = Saturation vapor pressure (monthly average), inches Hg.

e_2 = Vapor flexibility at a height of 2 meters.

F = Leakage.

F = Rate of mass infiltration at time t.

F = Total infiltration.

f (u) = A function in terms of wind speed at a some standard height.

f = Infiltration capacity (percolation) for time t.

f_c = Constant rate of infiltration after long wetting.

f_o = Initial infiltration capacity.

f_c = Final infiltration capacity = apparent saturated conductivity.

f = Natural recharge (rainfall – transpiration, surface runoff & infiltration).

F = Loss rate , mm per hour.

f = Possible head loss.

GI = Growth index of crop, percent of maturity.

H = Ponding depth.

H = Equivalent evaporation of total radiation on surface of plants.

H = Depth of confined aquifer

H = Depth of groundwater basin (m)

H = Saturated thickness of aquifer, m

H = Difference in elevation between two points after subtracting projections, feet.

h = Relative humidity, %

h = Depth below original water level.

h = Stage (gauge height).

I = Surface inflow.

I = Input (volume /time).

I = Part obstructed or trapped from rainfall.

I = Average intensity of rainfall, mm/hr.

I = Rainfall intensity, cm/hour.

I = Inflow to reach.

I = Rate of inflow, m^3/s.

I_t = Index value at t days later, mm.

I_o = Initial value to the index, mm.

i = Rainfall intensity, in or mm per hour.

i = Groundwater flow through area under consideration.

i = Hydraulic gradient.

K = Dimensionless constant.

K= Coefficient, storage constant (s), slope of relationship of storage-weighted discharge relation.

K_o = Loss coefficient at start of storm.

K_s = Effective hydraulic conductivity.

k = A recession constant.

k = Empirical constant for rate of decrease in infiltration capacity.

k = Proportionality factor, coefficient of permeability, or hydraulic conductivity (has dimensions of velocity).

k = Coefficient of permeability of the aquifer, m/s.

k = Constant of proportionality = reciprocal of slope of storage curve.

L = Latent heat of evaporation.

L = Latent heat per mole of changing state of water.

L = Specific heat of the vapor.

L = Latent heat of evaporation of water.

L = Accumulated loss during the storm, in or mm.

l = Distance in direction of stream line.

L = Depth to wetting front.

L = Maximum distance to entry point.

L = Length of stream, km.

L = Length from remotest point in basin to the outlet.

L = Main stream distance from outlet to divide.

l_{ca} = Stream distance from outlet to point opposite to basin centroid.

M_o = Average annual flow, m^3/s.

mcs = Water content at saturation that equals porosity (water content at residual air saturation).

mc = Water content at any instant.

N = Normal annual rainfall, mm.

N = Occurrence frequency, usually estimated once every N years = 10/n.

N = Net water infiltration rate resulting from rainfall.

N = Days of the month.

n = Number of days of the month (days of freezing not counted).

n = Number of occurrences in 10 years.

n = Retain coefficient equivalent to coefficient of friction.

n = Number of stations.

n/D = Cloudness ratio.

ne = Effective porosity, dimensionless.

O = Output, volume/time.

o = Groundwater outside borders of region.

P = Total pressure of humid air.

P '= Pressure of dry air.

P = Precipitation during a specific time period, year water.

P = Amount of rainfall, mm.

P = Amount of precipitation (rainfall).

P = Rainfall.

P = Total storm rainfall.

P = Atmospheric pressure, mbar.

P_e = Net rain representing portion of rainfall that reaches water courses for direct surface runoff.

P '= Decrease in rain.

P_x = Missing measurement or inaccurate record from station x, mm.

P_{av} = Arithmetic mean of rain, mm.

P_i = Amount of rainfall in station I, mm.

P_{mean} = Average rainfall in the region, mm.

P_i = Record of rainfall in station I, mm.

$\overline{P}$ = Average rainfall depth over the area.

P = Point rainfall depth measured at centre of the area.

Q = Total runoff (surface and ground) in basin during allotted time.

Q = Flow rate, flow

Q = Maximum rate of runoff, m3/s.

O = Surface outflow.

Q = Water flow rate.

Q = Total storm (surface) runoff.

Q = Steady state discharge from well

Q = Discharge as a function of time t, which changes from time t_1 to time t_2

Q_i = Average mean daily flow during a month i.

Q = Average annual flow.

Q = Peak discharge estimation expected to occur after heavy rains in the catchment basin, L/s.

Q = River flow.

Q_p = Discharge by probability of P , m^3/s.

Qa = Actual discharge (measured), m^3/s

Q = Steady state discharge (discharge from rating curve).

Q_t = Discharge at end of time t.

Q_a = Discharge at start of period.

Q = Discharge.

q = Flow rate of injected tracer to stream.

q = Ground water production from wells and disposal channels.

q = Flow in aquifer per unit width of basin, m^3/s/m.

q_p = Peak flow.

R = Bowen ratio.

R = Gas constant, L×atmosphere/K.

17

R = Radiation balance

R_A = Agot's value of solar radiation arriving at atmosphere.

R_1 and R_2 = Working values, representing indices of storage.
r = Addition and increase of underground water due to egress of
    surface water.
r = Distance from well.

r_a= Aerodynamics resistance.

r_H = Hydraulic radius, m.
r_s = Net physiological resistance.
S = Slope of energy line.
S = Storage coefficient.
S = Change in storage = Total precipitation over collecting surface.
S = Effective surface retention.
S = Volume of water, storage.

S = Stored volume of surface and ground for specified time period.
S = Change in reserved moisture in river basin.
S = Available storage in surface layer.
S = Absolute slope, m/m.

S = Stream slope, %
S = Steady stage energy gradient at the time of measurement, m/m.
S = Storage, m^3

S_d = Storage in low-lying areas.

S_f = Suction (capillary) head at the wetting front.
s = Increase in storage.
s = Drawdown.
T = Absolute temperature, K.

T = Transmissivity, m^2/day.

T = Number of seconds in a year (86400 = the number of seconds
        per day).

T = Base time (time from beginning and end of the flood).
T = Tme base for unit hydrograph, day.
T_r = Residence time.
Tw = Water surface temperature, °C.

Ta = Air temperature, °C.
t = Average dry temperature of the month, ° C.
t = Dry temperature.
t = Rainfall duration, minutes.

t = Time, day.

t_p = Basin lag, hours.

t_p = Time lag in the basin, hours.
t_c = Time of entry, minutes.
t_r = Unit rain period (duration of unit hydrograph).
t_{pR} = Basin lag for a storm of duration t_R, hours.
t_w = Temperature of wet bulb temperature (humid temperature).
$(t*)$ = Gamma inverse function for storm time.
tf = Total time during which rainfall intensity is greater than W.
t_o = Temperature of evaporation surface.
t_2 = Air temperature at a height of 2 meters.
V = Amount of moisture leaking for the period.

V = Volume of runoff.
V_f = Final molar volume for states of liquidity and gaseous.
V_i = Primary molar volume for states of liquidity and gaseous.

v = Flow velocity, m/s.
v = Wind speed, m/w.
v = Velocity of water flow (= specific velocity).
v' = Average pore velocity (actual or real velocity), m/s.
v = Relative (specific) velocity, m/s.
v = Specific velocity in the horizontal x direction, m.
V_s = Surface velocity.
V_{av} = Average velocity in a section of flow.
U = Groundwater inflows and outflows.
$U - O$ = Quantity of flow in river (surface and underground).
U = Velocity of flood wave (flood wave celerity), m/s
u = Wind speed, miles/hour.
u = Wind speed in the region at height z_o, m /s.
u_o = Wind speed at least height (at anemometer) z_o, , m/s.
u_1 = Wind speed at a height of 1 meter, m/s.
u_2 = Median value of wind speed at a height of 2 m above water
        surface, m/s.
u_2 = Wind speed at a height of 2 meters, m/s.

u_2 = Wind speed at height of 2 m, m/s.
u_6 = Wind velocity at a height of 6 meters above the surface, m/s.

$u (T, t)$ = Ordinate of the T-hour unit hydrograph derived from those
        of the t hour unit hydrograph.

W = W–index = Average infiltration rate during the time rainfall intensity exceeds the capacity rate.

W = Upper limit of the value of soil humidity that do not move by capillary action from the soil to the surface, %

W_v = Volumetric soil moisture, %

X = Dimensionless constant for a certain river reach, dimensionless.

x = Distance along flow line, m.

x, k = Constants derived from observed portion of curve.

y_o, y_n = Flows at beginning and end of flood, respectively.

Y_1, y_3 = Odd flows.

Y_4, y_6 = Even flows.

z = Height, m.

z_o = Height of anemometer (minimum height), m

α = Constant.

α = Coefficient of aquifer

γ = Psychrometer constant.

γ = Proportionality factor.

γ = Psychrometer constant fixed device to measure humidity.

Δh = Change in humidity for the period.

Δt = Difference between evaporation surface and air temperature.

ρ = Density of water

θ = Psychometric difference, ° C.

Δ =-Slope of vapor pressure curve at t = tan α

$\Delta s/\Delta t$ = Rate of change in reach storage with respect to time.

Δt = Routing Period.

ρ_a = Air density.

ρ_w = Density of water.

ϕ = Head loss over an appropriate base level.

ϕ = Potential head loss, m.

Chapter One

Computer models: Theory and Practice

1.1 Introduction and general design

Many decades ago, computer modeling was considered a black art, known only by a few, and mastered by even a fewer scholars. Today, with the information explosion that happened (as *is* happening) over recent years, the landscape of computer modeling has greatly changed. Computers span a wide spectrum of devices (some of which are not even perceived as computing devices, though they are): starting at a conventional full-sized desktop computer with all its ornaments. Running through laptops, notebooks, tablets, smartphones, and even smart-watches and glasses and embedded systems in a lengthy list, computational theory and practice incorporate almost *any* device in use today. Some of these devices are general in purpose; some are specialized, even highly specialized like an embedded Linux or Windows operating system running on a modern wrist-watch, or that operating a sophisticated MRI machine in a busy hospital or health center.

Using computer in engineering modeling is a common practice nowadays. Thanks to the widespread use of computers, and the popularity of high level programming languages, what was once an elite science is now a hand-reach away. This has become so true to the extent that practicing professional engineering without working knowledge of, at least, the basic computer modeling methods is, to say the least, considered a major drawback in practical advancement.

There are multiple computer models, with different implementation methods, which differ according to the:

1. underlying programming language used (be it a general purpose or a specialized engineering - or else - programming language),
2. computer platform (most commonly used in market being the Intel-compatible x86 or AMD64 architectures),
3. operating system workings (again, common OSes in the market including – but not limited to – MS-Windows, Linux with all its flavors, MAC OS, UNIX, among others), and, above all,
4. engineering backbone used in modular design.

As such, using computers in modeling is not an easy task. The modeler/programmer has many angels of view to consider in model design:

- What kind of architecture is the model intended to run at? Is it for desktop PCs? For embedded operating systems? To be run under 32-bit or 64-bit compatible processors? To work under single-processing or multi-processing environment? ... etc. Most general purpose engineering models, being common and easy to run with no special hardware needs, will be run under one or another Intel-compatible processor, running on an IBM-compatible PC or laptop, under 64-bit (supporting 32-bit) single or multi-processing environment (not to mention multithreading programs).
- What kind of operating system is it going to support? This depends not only on the programmer's preference and experience, but also is enforced by the audience who will run and use the software. There is no point in writing Windows native executable programs for a company that runs Linux based computers, for example. This specific example may look superficial, but there are many other examples in living programming modeling that are much more complicated, but the same idea holds.
- Based on the last point, cross-platform support is becoming an important issue in software engineering. To reach the widest possible group of audience, programmers must

consider writing code to be run under the largest possible (or at least the most common) types of operating systems. This can be achieved by writing code in a cross-platform language that is run by an interpreter or an on-the-fly compiler (like Java or the .NET framework languages). The hard way to go is to write the code for one OS, migrate it (which needs code revision and/or editing) to another platform, using OS-native tools to recompile and rebuild the program. This way is harder, takes longer time, is error-prone, and every change in code will need to be mirrored into all other versions, but is guaranteed to end with executables that are native to each operating system, are faster and more reliable.

- What programming language is going to be used? Some general programs need general purpose languages like VB, VC++, C/C++ among others. Programming for cross-platforming may involve other languages like Java or Qt/C++. Graphical detailing, mathematical modeling, web programming, database access, all need special purpose programming languages that may, or may not, need their specialized compilers/interpreters to run effectively.

Generally, the programmer will need to put an overall blueprint for the program flow, which is commonly known as the program's algorithm. This will include:

- Intended purpose of the program, such as "A program to calculate the square root of a number"
- Kind of inputs the program accepts e.g., "program accepts one floating point number"
- Steps of computations needed to be done, including any equations or special mathematical formulae that will be used, e.g., "calculate the square root using sqrt function"
- Outputs expected from the program, e.g., "prints the sqrt of the input"

This may, or may not, be supported by a visual layout or a "flow chart". For the sake of simplicity, all programs in this book included only the textual representation of each program's algorithm.

23

1.2 Programming examples

This book is not intended to be an introduction to computer programming (although a crash course is given herein to start with), nor is it going to dive into details about computer architecture or operating system internal workings. The reader is assumed to have a basic working knowledge in the following areas to be able to read, follow, test, and implement the programming examples given through this textbook:

- Basic computer knowledge and how-to's (as how to open/operate/setup a computer workstation),
- Installing and running any operating system (computer programs in the book are programmed using Visual Basic 10 under Visual Studio 2010 Professional Edition, tested and run on an MS-Windows XP and 7 workstations. It can be tested on Fedora Linux 20 system using WINE – WINdows Emulator – from www.winehq.org, to run the executables. The source code files can be viewed in any text editor if the user is using an OS other than Windows (or using Windows without VS installed), but to manipulate and rebuild the programs the user will need other workarounds, like using MonoDevelop from www.go-mono.com/mono-downloads/download.html, for example, as an IDE that can run and compile .NET framework programs). The programs should run on any system that supports EXE file formats and a .NET framework JIT compiler (Just-In-Time compiler), or a similar software to run the CIL (Common-Intermediate Language) executables.
- At least basic programming knowledge with Visual Basic is required. ANonetheless, the example programs are straightforward and the language structure is clean enough to make it easy for programmers experienced with other programming languages to follow program logic and run/test the programs.
- Microsoft Visual Studio 2010 Enterprise Edition was used to program, compile, and debug the examples presented in this book. It can be obtained from www.visualstudio.com/en-us/products/visual-studio-express-vs.aspx (for the express edition, a lightweight edition available for the programming

languages separately), or from go.microsoft.com/fwlink/? linkid=240162, or getting the software CD/DVD from your software vendor, or by simply looking for "Microsoft Visual Studio Download" in Google search engine.

As is commonplace today for GUI (Graphical User Interface) programs written for windowing systems, the programs have two parts: the GUI design (what the user sees, which is usually translated to a code file with special syntax by the designer or IDE), and the code doing the real work behind the scenes. The code for each program in the book is included in the text along its respective example. The code provided herein goes into the main program window source file (usually named *Form1.vb* by default, if not indicated otherwise in the text), and as all the programs here are designed to be single-windowed for the sake of simplicity, this will be the case. As for the GUI design part, all the user interfaces are included as snapshots in Appendix (G) in the back of the book. The user can refer to them when reading/testing the programs to link the controls used in the design with the code (as the default names assigned to the controls are retained, that is, TextBox1, Label2, Form1, and so on, are left unchanged, to make it easy for the reader to follow what-goes-where).

The programming code was designed to be at the minimum level needed to perform the tasks required, at the same time not to be too short, scrambled, or ambiguous, so that new and novice programmers can follow with ease. Precise, concise code is a good way to write computer modules. This is not always the case. For example, if a certain function can be done in two ways, one which is two instructions long, the other is eight, there are generally two approaches: either use the short version, but include detailed (even lengthy) comments – remember comments doesn't get to the final executable, so not to worry about final code size. The other way is to use the lengthy operation, trading size and speed for code clarity. Consequently, when returning to the code a few months later, one can still understand what he/she did that late night when he/she wrote that piece of code (and again, thorough comments in the code will help this).

For each task performed in the presented examples, the reader will find prompt documentation through the source code in the relevant parts to ease following the code and make it a rewarding and productive task. This is especially if a certain function performs a lengthy operation or a specific programming concept is felt to be complex or advanced,

1.3 The .NET Framework and Visual Basic.NET

The .NET Fx (pronounced "dot net framework") framework is a large group of functions. These functions serve to do virtually whatever a programmer wants the system to do. That code is grouped into classes, with each class containing functions that work on certain targets to do certain things. For example, the Drawing class holds the functionality of drawing lines, rectangles, circles, coloring and much more.

By abstracting the programming language from what really happens under the hood, a new concept of programming is introduced. The programmer is not bound to learning specific languages to do complex tasks that can't be done in other languages, as it once used to be. For example, VB was called a 'sandboxed' language, because it protected the programmer from knowing the lower-level functionality of the operating system. With the new programming concept of the .NET Fx, every program that is written in a .NET-compatible language (C# - pronounced 'see sharp' – the *de facto* .NET language, C++, J++ and others). will be compiled to an intermediate language, known as the Common Intermediate Language (CIL). One can think of this as akin to the assembly language, but specific to the .NET framework, not to the underlying computer architecture. All .NET programs translate to the CIL, which will finally be compiled into the final executable file to be run by the machine. At run time, a special program, part of the .NET Fx toolset, known as the 'Just-In-Time Translator' - JIT, loads the program to be run into the memory, reads in the program instruction by instruction, and tells the processor what to do in processor-specific instructions.

Visual Basic.NET was selected as a programming language for the book's examples for many reasons that incorporate:

- MS-Windows is one of the most (if not *the* most) popular operating system in use today. Chances are, when buying a new PC or laptop (or even a mobile phone, tablet or iPad), the user will face this OS first and foremost.
- .NET Fx is becoming so popular as a dominant programming platform in computer industry, that it will be counterproductive not to learn and understand how to implement engineering models using this platform.
- Visual Basic.NET came from the old good beginners' friend BASIC, which was infamous for being an easy to learn language, albeit with moderate capabilities, but above all, having so many layers of abstraction that the programmer doesn't need to bother about the low-level workings of the computer, he/she just focuses on programming. For this, BASIC, and its descendent VB, which then evolved to VB.NET, is considered one of the easiest programming languages to learn, especially for beginners.

Visual Basic.NET is not a for-all-purposes language. Praise it enough for good abstraction, clear structure, and educational appeal, the language is a property of Microsoft Corporation. Being a proprietary software, along with the .NET framework itself, it is not easy (not impossible though) to directly use the language to program native software under other operating systems (for example Linux, which is becoming an important player in the arena of operating systems nowadays). Some workarounds includes using OS emulators, or .NET-compatible IDEs and compilers (discussed above). Sometimes it is just as easy to use another programming language like Java or Qt/C++ that supports cross-platforming out-of-the-box, pros and cons for this scheme are discussed above.

As much as you are encouraged to explore (and implement) cross-platform model design and, it was not used here because it would add so much overhead to the discussion that it will stray the reader out of the main point of the day: basic model design. That is, the basic function of the program examples in the book is to explain model designs in the most clear and straightforward way, not to make the reader the next VB expert. Therefore, programmers in other languages are encouraged to translate the example programs to their language of expertise, and to cross-reference with the code example herein, it ought to be a productive and educational task. You are free to use the code and manipulate it as freedom is, as long as it is not used in a commercial-level program, in which case the programmer is kindly asked to reference the source text in their programs.

1.4 Working with Visual Studio.NET

In this section, a quick hands-on tutorial will be presented to help novice readers get aboard the Visual Studio.NET ship. The first step is to setup Microsoft Visual Studio. It can be obtained from one of the sources listed above. After downloading and installing the software, run it from the Start menu. After it finishes loading, the screen will be like the one presented in figure 1.1.

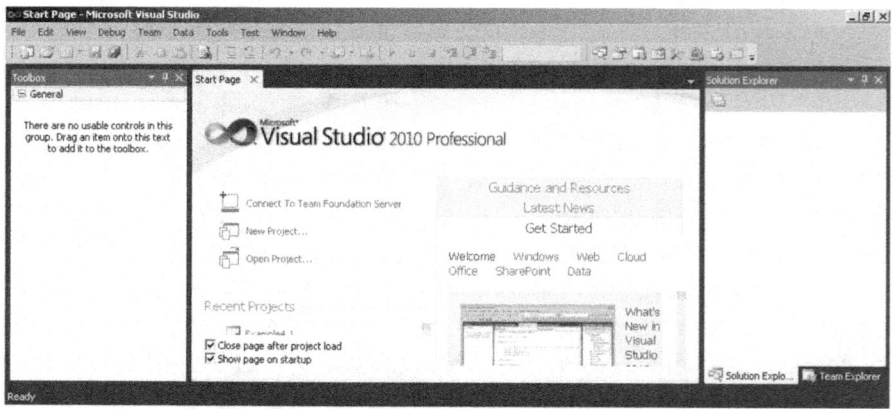

Fig. 1.1 Visual Studio screen.

Select File → New Project. Enter the name in the Project name (Name field, bottom of window), leave other options as is, and click OK. The following screen will be like the one shown in figure 1.2.

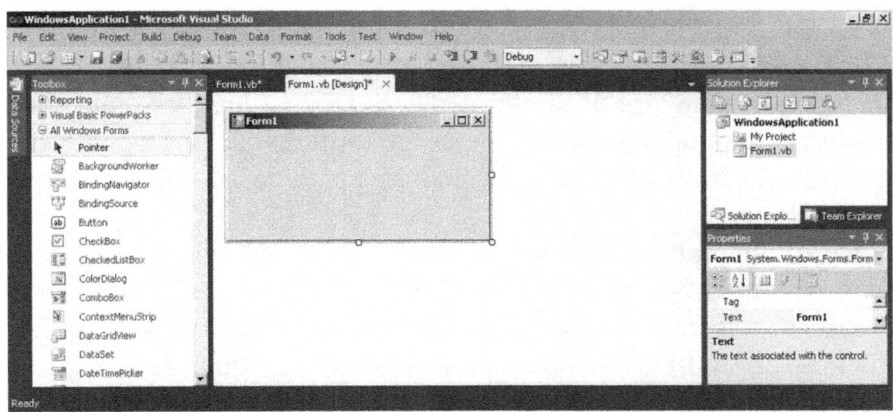

Figure 1.2 New Project Screen.

This is the form design view. On the lefthand is the Toolbox. Here are the controls that can be added to the design. Search down for *Label,* Double-click on it to add it (or click on it and drag it to the form). Reposition it on the form by holding the left mouse button and moving it, until it looks similar to the one depicted in figure 1.3.

Figure 1.3 Label Screen.

Go ahead and add a button, repositioning it below the label.

Click on an empty area on the form. Small boxes will appear around the corners and sides of the form. These are call *handles*. Select the handle in the middle-bottom side of the form by clicking on it. Keep the button down and move the mouse up to resize the form to a smaller size.

Click on the button to select it. On the righthand side is a window named *Properties*. Here you will find properties of the selected control (as text, size, color …). Search for property *Text*, double click on it, and enter: "Say Hello", without the quotes, then press Enter. The text on the button will change to *Say Hello*. Resize the button by using the handles to make it wider.

Click on the form in an empty area, select the *Text* property in the properties window, and change it to: *Hello program*. The end result should look comparable to the one portrayed in figure 1.4.

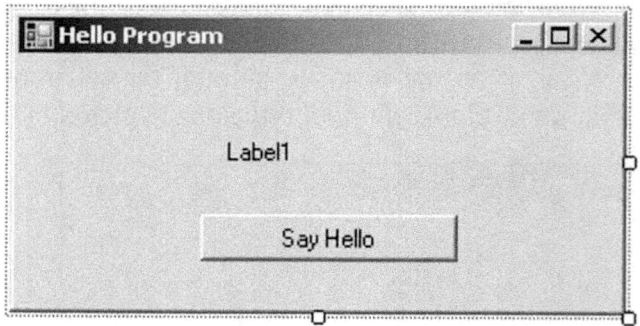

Figure 1.4 Hello program Screen.

To add some code, double click on the form in an empty area. The code view will open. It will look like:

```
Public Class Form1

Private Sub Form1_Load(ByVal sender As System.Object,
ByVal e As System.EventArgs)...

    End Sub
End Class
```

Position the cursor (by arrow keys or mouse click) in the empty line between line starting   'Private Sub Form1_Load' and line 'End Sub'. Type in:

```
Label1.Text = ""
```

This will erase the text 'Label1' from the label when the form loads. Why not do it in design view? Because the label text will be empty and it will be hard to find it on the form.
To go back to design view, select View → Designer, or SHIFT+F7, or select its icon from the Solution Explorer window on the righthand side.
Now double click on the button, it will re-open the code view, this time adding another function stub below the first one, but also included between the Class Form1 and End Class statements:

```
Private Sub Button1_Click(ByVal sender As System.Object,
ByVal e As System.EventArgs)...

End Sub
```

This will handle the event button_click, which happens when user clicks on the button. Position the cursor between 'Private Sub Button1_Click...' and 'End Sub' and type:

```
MsgBox("Hello World!", vbOkOnly, "Say Hello Program")
Label1.Text = "Thank you for using the Hello world."
```

Now the program is ready. To make executable, there are two modes: Debug and Release. Debug is good when developing the program; it includes debugging information in the executable to make tracking bugs easier. When you finish the development and want to ship the project to the world, build it in Release mode. Select Debug → Start Debugging, or press F5, or click the green arrow on the toolstrip. The program will run. It will show the window. Click the button, it will show a message box with OK button, wen you click it the

31

message box will disappear and the label on the form will show the message you entered above. When finished, click the red X on the right upper corner of the window, press ALT+F4, or select Stop Debugging (appears as a blue box) from the toolstrip.

To change Debug and Release modes, change the select from the toolstrip as shown like the one represented in figure 1.5.

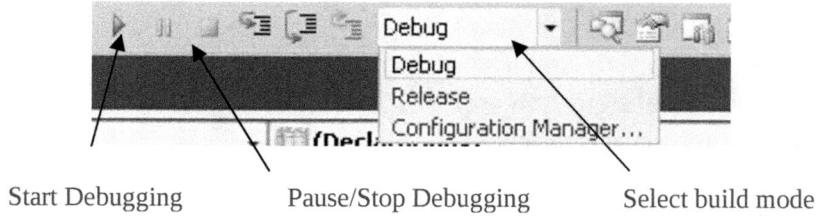

Start Debugging   Pause/Stop Debugging   Select build mode

Figure 1.5 Changing Debug and Release modes Screen.

To save the project, press CTRL+SHIFT+S, select File → Save All, or press the three blue diskettes on the toolstrip. Visual Studio will show a dialog box like the one revealed in figure 1.6.

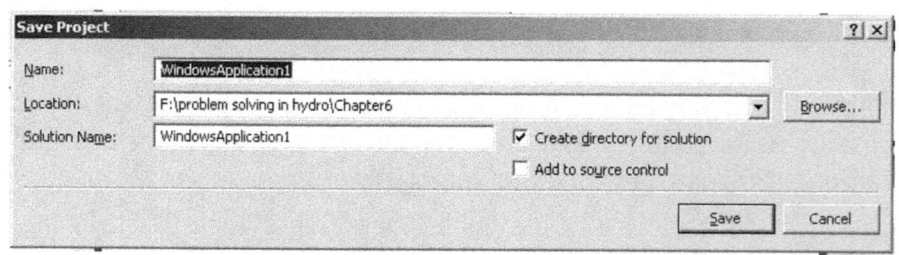

Figure 1.6 Save dialog box Screen.

Enter the project name in the first field. Enter the folder to save under in the second field; you can browse for a folder by clicking on Browse. You can change the Solution Name field or leave it as is which the project Name is, by default. Notice the *Create directory for solution* is selected, which means VS will make a directory under the *Location* you selected. For example if Location is *D:\Docs*, then VS will create *D:\Docs\WindowsApplication1* and place the project files in it.

Subsequent saves will not show this dialog box, it is shown the first time a project is saved.

If you open the project directory in Windows Explorer, you will find the following files and dirs, assuming your project name is *WindowsApplication1* (not all are listed):

- *WindowsApplication1.sln:* this is the solution file. When opening your project after closing VS, you will double click this file to reopen the project.
- *WindowsApplication1 (Folder):* This contains the files and resources of the project's components. Enter the folder you will find:
    1. *Form1.vb:* code you entered for Form1 controls (button and form event handlers)
    2. *Form1.designer.vb:* code generated by the designer to describe the layout of the form you designed
    3. *Form1.resx:* resources associated with the form, such as images or icons
    4. *bin:* This folder contains two folders: *Debug* and *Release.* According to the build mode you selected, the project files will be under one (or both) of these directories. Inside either one you will find five (or more) files, the one name *WindowsApplication1.exe* is the program executable, you can double click it to test it as the normal user will usually do to your application

1.5 Visual Basic.NET programming primer

Visual Basic.NET has quite a handful of instructions, all of which can not be discussed here. Instead, the more basic and commonly used instructions are discussed. Advanced programmers can skip through this section, although they are encouraged to review the table summarizing the commoner 'VB.NET functions used throughout the book.

Variables and constants. Data can be used hardcoded as constants and indirectly as variables. A syntax like:

```
Const G = 9.81      'Gravitational acceleration
```

defines a value that will not change through program execution. In fact, the compiler will place the value 9.81 whenever it finds the name G in the code, so the final executable will only contain the value.

The types of variables are:
Numbers can be positive, negative. They are expressed as integer quantities (with no fractal part) or floating point quantities (can be single or double precision, occupying upto 64 bits in storage). Integers are declared as Short, Integer or Long, with a capacity to store 2, 4, and 8 bytes of memory respectively. That means Short integers can have values in the range -32768 to +32767.
Define a variable as:

```
Dim var as Integer        'Defines an empty integer variable
Dim X as Double = 1.234  'Defines a floating point variable
                          'and initializes it
```

Strings and chars define a sequence of characters and one character, respectively. The value is written between single or double quotations.
Booleans accepts only one of two values: True or False. It is heavily used in logical comparisons to direct program flow through loops and jumps.

Operators indicate arithmetic operations:
+   addition
-   subtraction
*   multiplication
/   division
^   exponentiation

Order of execution gives priority to operands inside parenthesis, then exponentiations, multiplication and division, with addition and subtraction to be carried out last. If two operands of the same priority exist in one formula, the order is left-to-right in execution.

To add a comment to the code, use the **REM** statement, or the single quote (‘). Comments are ignored by the compiler.

The **END** statement is used to indicate the end of a function or a subroutine, use as *'End Sub'* or *'End Function'*. If exit needs to be done in the middle of executing some loop or function, use *'Exit'* instead:

```
For I=1 to 10
    IF I>J THEN Exit For
Next
```

The **GOTO** statement is used to transfer control to another part of the program. It can be combined with 'IF' clauses to bind the branching decisions to different situations, making it conditional branching:

```
IF I=J THEN
    GOTO Equals
ELSE
    GOTO NotEquals
END IF
…
Equals:
…
NotEquals:
…
```

Engineering programs need to access math functions frequently. Most of the mathematical functions are grouped in the namespace 'System.Math'. (A namespace is a name used to define a group of classes that are related to each other in certain ways. The namespace can contain class definitions, constants (such as 'Pi' in the 'Math' namespace), and functions (such as the mathematical functions described below)). Some of the commonly used namespaces are: System.Math, and Microsoft.Windows.Forms. If a function or a member of a namespace is required, the namespace will need to be 'imported' or inserted at the top of the code, before the 'Public Class Form' statement, such as:

```
Imports System.Math
…
Public Class Form1
…(Do something using System.Math functions)
End Class
```

If the namespace is not accessed frequently in the code, just append the name of the namespace to the function required, as:

```
DIM X, LogX as Double
X = 10
LogX = Math.Log(X)
```

The IF-THEN statement is used to perform a conditional branching operation. The condition occurring between the IF and the THEN is a Boolean expression (see Table 1.2). It can be used in a block as follows:

```
IF {condition} THEN
    (What to do if condition is TRUE)
ELSE
    (What to do if condition is FALSE)
END IF
```

Table 1.1 summarizes the Boolean expressions.

TABLE 1.1 The Boolean Expressions

Operator	Relationship	Example*
=	Equal	X = Y is FALSE
<>	Not equal	X <> Y is TRUE
<	Less than	X < Y is TRUE
>	Greater than	X > Y is FALSE
<=	Less than or equal to	X <= Y is TRUE
>=	Greater than or equal to	X >= Y is FALSE
Not	Negates the following expression	NOT(X = Y) is TRUE

* Using X=1, Y=2 for demonstration.

36

To select an option from a group, the **CASE** construct can be used as:

```
SELECT CASE x
CASE a
    What to do
CASE b
    What else
...
CASE ELSE
    Default case if none is applicable
END SELECT
```

To execute some code for a predefined number of times, use the FOR TO loop. The statement needs a variable whose values change each time the loop is executed.

The DO...LOOP construct is similar to for, except it doesn't need to know how many times to execute the code. It needs a Boolean expression that is evaluated every time the loop is executed. The loop

```
DO WHILE X<Y
      ...Some code
LOOP
```

executes **as long as** the condition is true. On the other hand, the loop

```
DO UNTIL X>=Y
    ...Some code
LOOP
```

does the same thing, but in the reverse: it executes **until** the condition becomes true.

Sometimes an output needs to be formatted in a certain way. This can be done by a group of functions. The *Format()* function formats a string for output in a certain way, and the *FormatNumber()* formats a number and returns result as a string. Some of the styles used are predefined in the function. It takes a string input and a style, then applies the style to the input string and gives the result.

Drawing functions:

a) The *DrawLine()* is a graphics function that draws a line on the graphics screen. The coordinates of the ends of the line and the drawing color must be provided.

b) The *DrawRectangle()* is also a graphics function that draws a rectangle or a box on the graphics screen. The coordinates must be provided, along with the color of the lines.

c) *DrawEllipse()* and *DrawArc()* are used to draw circles, ellipses, and arcs, provided the X/Y coordinates, and the width and height are given. Also needs the color of drawing to be provided.

d) *DrawString()* draws a string text to the control, using the Font selected, with the specified color.

The drawing functions need a drawing board to draw on, usually known as a *Graphics* object. It is associated with controls such as a form or a PictureBox (to draw on a form or a PictureBox, respectively). The *Graphics* object must be created first, and then used to draw. For example:

```
Dim g As Graphics
g = PictureBox1.CreateGraphics
g.DrawLine(Pens.Black, 0, 0, 10, 10) 'The color, X,
                                     'Y, width,
                          'height
g.DrawString("Hello", New Font(FontFamily.
        GenericMonospace, 8),
        Brushes.Blue, 50, 50)
```

The **DIM** (dimension) statement is used to declare variables and arrays (An array represents a collection of related variables in memory, representing a list or table). It allows specifying the maximum number of elements for an array. Arrays are zero-based, so first element is at index zero, and the array's size is usually one-size more than the given dimension, for example:

```
DIM x(3) as Double
DIM y(3, 5) as Integer
DIM z() as String
```

38

The first statement defines x as a one-dimensional array with 4 double elements (0 to 3). The second defines y as a two-dimensional array with size 4 times 6, or 24 integer elements. The last one defines z as an array of strings, with no size specified. To use it, it will need to be resized later before access, for example:

```
ReDim z(4)
```

When programs grow bigger and more complex, they need to be fragmented to ease reading, and to group code that does certain functions together away from code that does other things. This is done using subroutines and functions. The **CALL** statement is used to transfer control to the subroutine, and the **END SUB** or **EXIT SUB** statement is used to cause control to transfer back to the part of the program that called it. The function is similar, except it must return a value to the caller. Some subs are predefined, for example when double-clicking on the form in design view, the VS IDE will automatically generate an empty sub called *Form_Load()*. To add user defined subs, enter in the code view (between the *Class Form1* and *End Class* statements):

```
Private Sub demoSub()
   …Some code
End Sub
```

Define a function with its return type (the value to be returned to the caller):

```
Private Function multiply(x as Double, y as Double)
                        as Double
   Return x*y
End function
```

Conveying messages to the user can be done using the **MSGBOX** and **INPUTBOX** functions. Those use a predefined dialog box, containing either a message with response options (MSGBOX with buttons of YES/NO or OK/CANCEL), or providing a means of simple input for the user (INPUTBOX). The parameters passed to the MSGBOX are: the message to the user (the prompt), the title (optional), and the buttons to display and/or the icon to show (a big 'I' for information, a red cross for error …).

Every control has properties that define how it looks and behaves. Some of the common properties are summarized in table 1.2.

Table 1.2 Some of the common control properties.

Property	Purpose	Example
Text	What is displayed on the control	TextBox1.**Text**="hello" Form1.**Text**="Example Program"
Visible	Show/hide the control	Me.**Visible**=True TextBox1.**Visible**=False
Checked	If the checkbox is checked	CheckBox1.**Checked**=True
Height and Width	Sets the size of the control	TextBox1.**Height**=20 TextBox1.**Width**=100
Multiline	Determine if a textbox supports multiple lines of output	textBox1.**Multiline** = True

To ease reading them, all programs in this book where written in a uniform way, with most of the calculations done in the *Button_Click()* event handler.. Most of the other assignments (giving names to labels, buttons, textboxes …) were written inside the code (the *Form_Load()* sub) to make the process of following the code easy. The default names of the controls were left unchanged, so the reader can follow easily between the code and the design (for example, *TextBox1, Label2* and *Button3* were left with their default names). Usually, in real-world programming, giving names to controls, setting the text of a form, resizing a button and other simple assignments will be done during design time using the Properties window in the VS IDE.

Table 1.3 presents a summary of common VB.NET statements and functions. More exhaustive lists can be found in dedicated programming books.

Table 1.3 Summary of common Visual Basic.NET Statements and Functions.

Statement/ function	Purpose	Example
ABS	Keeps absolute value	Y = ABS(x)*
ATAN	Returns arctangent	Y = ATAN(x)*
CALL	Calls a subroutine	CALL IHR
DrawEllipse, DrawArc	Draws a circle, an ellipse, or an arc	g.DrawEllipse(Pens.Black,10,10,40,40)
DrawLine	Draws lines	g.DrawLine(Pens.Black, 0, 0, 10, 10)
DrawRectangle	Draws a rectangle or a box	g.DrawRectangle(Pens.Black, 0, 0, 20, 40)
DrawString	Draws a text string, given the brush color, the X-Y coordinates of where to start drawing, the string to be written, and the font to be used	g.DrawString("Hello", New Font(FontFamily.GenericMonospace, 8), Brushes.Blue, 50, 50)
COS	Determines cosine function	Y = COS(x)*
DIM	Defines variables or arrays	DIM x as Integer DIM y(5,200)
END	Ends a function or subroutine	END SUB END FUNCTION
EXP	Determines exponential function	Y = EXP(x)*
FOR and NEXT	Defines the beginning and end of a FOR-TO loop	FOR m = 1 to 39 NEXT m
FormatNumber	Formats number in specific way and returns result as string	X = FormatNumber(3.141, 2) 'Returns 3.14

GOTO	Transfers program to a remote statement	GOTO 3000

IF-THEN	Conditional run	IF y<8 THEN
		ELSE
		END IF

LOG	Returns the natural logarithm (to base e)	Y = LOG(x)*
LOG10	Returns logarithm to base 10	Y = LOG10(x)*
REM	Inserts comments and remarks in the program	REM Program listing
SIN	Determines the sine function	Y = SIN(x)*
SQRT	Determines square root of a value	Y = SQRT(x)*
ToString	Converts item to string	X = 5 Str = X.ToString
ATAN	Determines the tangent	Y = ATAN(x)*
VAL	Changes a string to a numerical value	Y = VAL(x)

* Requires import of the *System.Math* namespace

1.6 Programs on the Accompanying CD/DVD

The programs on the CD accompanying this book are all source and executable forms. They are grouped in folders named by chapter. A separate folder including the executables is named "EXE Alone". Simply, inserting the CD will auto-run it. If not, open it in MyComputer and run HYDROLOGY.exe, which is a self-extracting archive. It will uncompress the folders mentioned above.

There should be no problem running the CD and examples programs, but for the sake of completion, the following system hardware and software specifications as required:

- A PC or a laptop.
- A CD/ROM drive.
- A hard disk drive with at least 30 MB free space (for the code and EXE files).

- A math co-processor chip is recommended to enhance performance (usually incorporated into modern Intel processors).
- A Windows operating system version XP, 2003, 7.
- An installed .NET framework package version 4.x or later (see the User's Manual on how to download and install the latest version).

For users using other operating systems, some solutions were described earlier (using emulators, etc...).

1.7 Where to go next...

Programming is a continuous process. It needs writing code, running tests, debugging, modifying code, rerunning, re-debugging, and so on, until the final program is ready to ship.

To become more professional and write better code, one needs to read more programming books, read code written by other (more experienced) programmers, and search on the internet. A good place to start with is the Microsoft Developer Network, http://msdn.microsoft.com. This is a vast library with code examples, discussions, and much more.

Chapter Two

Meteorological data (Humidity, temperature, radiation and wind)

2.1 Humidity

Maximum value of water vapor that can exit in any one space is a function of temperature, and is practically independent of the coexistence of other gases. When maximum amount of water vapor (for a given temperature) is contained in a given space, the space is said to be saturated. The pressure exerted by a vapor in a saturated space is called "saturation vapor pressure. "

2.2 Relative humidity

Relative humidity is the percentage of actual vapor pressure to saturation vapor pressure. It is the ratio of amount of moisture in a given space to the amount a space could contain if saturated.

2.3 Dew point

Dew point denotes the temperature at which space becomes saturated when air is cooled under constant pressure and with constant water vapor pressure. Also dew point is the temperature having a saturation vapor pressure, e_s equal to the existing vapor pressure, e.

Example 2.1

An air mass is at a temperature of $20°$ C with relative humidity of 75%. Using the following equation

$$h = 100 \; x \; \frac{e}{e_s}$$

find: saturation vapor pressure, actual vapor pressure, the deficit in saturation and dew point. (B.Sc., DU, 2011)

Solution

a) Data given: T = $20°$ C, h = 75 %
b) Find from tables (see annex) the value of saturated vapor pressure at a temperature of $20°$ C,
    Value of saturated vapor pressure e_s = 17.53 mm Hg
c) Substitute data in humidity equation:

$$h = 100 \; x \; \frac{e}{e_s}$$

Where:
h = Relative humidity, %, which describes the ability of air to absorb additional moisture at a given temperature.
e = Actual vapor pressure
e_s = Saturation vapor pressure. (Numerical value of e_s are to be found from tables).

h = 100 * e/e_s, find real vapor pressure:
75 = 100 × e ÷ 17.53
Real vapor pressure e = 13.1475 mm Hg.

d) Find saturation deficit as follows:
    Saturation deficit: e - e_s = 17.53 - 13.1475 = 4.3825 mm Hg.
e) Find dew point as the temperature at which the values of e_s and e are equal.
    Since: e_s = 13.1475 mm Hg, then one can find the temperature, and then the dew point can be found from tables and for the value of e_s = 13.1475, dew point = 15.4 $°$ C.

Program 2.1 Listing:

```
'********************
'EXAMPLE 2.1
'********************
Public Class Form1
    Dim es(30) As Double

    Private Sub Form1_Load(ByVal sender
         As System.Object, ByVal e
      As System.EventArgs) Handles MyBase.Load

      Me.Text = "Example 2.1"
      Me.FormBorderStyle =
      Windows.Forms.FormBorderStyle.FixedSingle
      Me.MaximizeBox = False

      Label1.Text = "Temp. (C)"
      Label2.Text = "Relative humidity (%)"
      Label3.Text = ""
      Button1.Text = "&Calculate"
      'setup vapor pressure table
      es(0) = 4.58
      es(1) = 4.92
      es(2) = 5.29
      es(3) = 5.68
      es(4) = 6.1
      es(5) = 6.54
      es(6) = 7.01
      es(7) = 7.51
      es(8) = 8.04
      es(9) = 8.61
      es(10) = 9.2
      es(11) = 9.84
      es(12) = 10.52
      es(13) = 11.23
      es(14) = 11.98
      es(15) = 12.78
      es(16) = 13.63
      es(17) = 14.53
      es(18) = 15.46
      es(19) = 16.46
      es(20) = 17.53
      es(21) = 18.65
      es(22) = 19.82
      es(23) = 21.05
      es(24) = 22.27
      es(25) = 23.75
      es(26) = 25.31
```

```
        es(27) = 26.74
        es(28) = 28.32
        es(29) = 30.03
        es(30) = 31.82
End Sub

Private Sub Button1_Click(ByVal sender As
    System.Object, ByVal e As
    System.EventArgs) Handles Button1.Click

    Dim _e, T, h As Double
    T = Val(TextBox1.Text)
    h = Val(TextBox2.Text)

    If T < 0 Or T > 30 Then
        MsgBox("Enter temp. value between 0
            and 30 inclusive.", _
                vbOKOnly, "Invalid temp.")
        Exit Sub
    End If

    _e = h * es(T) / 100
    Label3.Text = "Real vapor pressure e = " +
        FormatNumber(_e, 2) + " mmHg"
    Label3.Text += vbCrLf
    Label3.Text += "Saturation deficit = " + _
            FormatNumber(Math.
            Abs(_e - es(T)), 2) + " mmHg"
End Sub
End Class
```

2.4 Water Budget

Hydrological cycle is a very complex series of processes.
Nonetheless, under certain well-defined conditions, the response of a
watershed to rainfall, infiltration, and evaporation can be calculated
if simple assumptions can be made.

Example 2.2

For a given month, a 121 ha lake has 0.43 m^3/s of inflow, 0.37 m^3/s
of outflow, and the total storage increase of 1.97 ha-m. A USGS
gauge next to the lake recorded a total of 3.3 cm precipitation for the
lake for the month. Assuming that infiltration loss is insignificant for
the lake, determine the evaporation loss, in cm over the lake for the
month.

Solution

Given,   P = 3.3 cm, I = 0.43 m³/s, Q = 0.37 m³/s, S = 1.97 ha-m,
Area, A = 121 ha
Solving the water balance for Inflow, I and Outflow, O in a lake
gives, for evaporation, E ;

$$\text{Inflow, I} = \frac{\left[0.43\,\frac{m^3}{s}\right]\left[1\,month \times \frac{30\,day}{1\,month}\right]\left[\frac{24\,hour}{1\,day}\right]\left[\frac{3600\,s}{1\,hour}\right]}{(121\,ha)\left[\frac{10000\,m^2}{1\,ha}\right]}$$

Inflow, I = 0.92 m = 92 cm

$$\text{Outflow, Q} = \frac{\left[0.37\,\frac{m^3}{s}\right]\left[1\,month \times \frac{30\,day}{1\,month}\right]\left[\frac{24\,hour}{1\,day}\right]\left[\frac{3600\,s}{1\,hour}\right]}{(121\,ha)\left[\frac{10000\,m^2}{1\,ha}\right]}$$

Outflow, Q = 0.79 m = 79 cm
<u>Additional data</u> :
Direct Precipitation, P   = 3.3 cm

Storage, S = $\dfrac{1.97\,ha.\,m}{121\,ha}$ = 0.0163 m   = 1.63cm

From water budget equation,
$P - R - G - E - T = \Delta S$

In this case,
$I + P - Q - E = \Delta S$
Therefore,
Evaporation, E :
E = I + P − Q - ΔS
E = 92 + 3.3 − 79 − 1.63
E = 14.67 cm

Program 2.2 Listing:

```
'*******************************
'EXAMPLE 2.2: Evaporation loss
'*******************************
Public Class Form1

    Private Sub Form1_Load(ByVal sender As
        System.Object, ByVal e As
        System.EventArgs) Handles MyBase.Load

        Me.Text = "Program 2.2"
        Me.FormBorderStyle =
        Windows.Forms.FormBorderStyle.FixedSingle
        Me.MaximizeBox = False

        Label1.Text = "Precipitation/month, P (cm)"
        Label2.Text = "Inflow, I (m3/s)"
        Label3.Text = "Outflow, Q (m3/s)"
        Label4.Text = "Storage increase, S (ha-m)"
        Label5.Text = "Area, A (ha)"
        Label6.Text = ""
        Button1.Text = "&Calculate"
    End Sub

    Private Sub Button1_Click(ByVal sender As
        System.Object, ByVal e As
        System.EventArgs) Handles Button1.Click
        Dim _E, P, Q, I, A, S As Double
        P = Val(TextBox1.Text)
        I = Val(TextBox2.Text)
        Q = Val(TextBox3.Text)
        S = Val(TextBox4.Text)
        A = Val(TextBox5.Text)

        I *= 30 * 24 * 3600
        I /= (A * 10000)
        I *= 100      'convert to cm
        Q *= 30 * 24 * 3600
        Q /= (A * 10000)
        Q *= 100      'convert to cm
        S /= A
        S *= 100      'convert to cm
        _E = I + P - Q - S
        Label6.Text = "Evaporation loss E = " +
            FormatNumber(_E, 2) + " cm"
    End Sub
End Class
```

Example 2.3

At a particular time, the storage in a river reach is 60×10^3 m³. At that time, the inflow into the reach is 10 m³/s and the outflow is 16 m³/s. After two hours, the inflow and the outflow are 18 m³/s and 20 m³/s respectively. Determine the change in storage during two hours period and the storage volume after two hours.

Solution

Given, I_1 = 10 m³/s
I_2 = 18 m³/s
O_1 = 16 m³/s
O_2 = 20 m³/s
S_1 = 60×10^3 m³
Δt = 2 hours x 60 min x 60 s   = 7200 s

1) Change in storage, ΔS during two hours :

$$I - O = \Delta S$$

$$\left(\frac{I_1 + I_2}{2}\right) - \left(\frac{O_1 + O_2}{2}\right) = \frac{\Delta S}{t}$$

$$\left(\frac{10+18}{2}\right) - \left(\frac{16+20}{2}\right) = \frac{\Delta S}{t}$$

Therefore,

$$\frac{\Delta S}{t} = -4 \frac{m^3}{s}$$

$$\Delta S = -4 \frac{m^3}{s} \times 7200\, s$$

$$\Delta S = -28800\, m^3$$

2) Storage Volume after two hours, S_2 :

$$\Delta S = S_2 - S_1$$

Rearrange,

S_2 = $\Delta S + S_1$
S_2 = $- 28\,800 + 60 \times 10^3$
S_2 = 31 200 m³

Program 2.3 Listing:

```vb
'*************************
'EXAMPLE 2.3: STORAGE
'*************************
Public Class Form1

    Private Sub Form1_Load(ByVal sender As
        System.Object, ByVal e As
        System.EventArgs) Handles MyBase.Load
        Me.Text = "Example 2.3"
        Me.FormBorderStyle =
        Windows.Forms.FormBorderStyle.FixedSingle
        Me.MaximizeBox = False

        Label1.Text = "Inflow, I1 (m3/s)"
        Label2.Text = "Inflow, I2 (m3/s)"
        Label3.Text = "Outflow, O1 (m3/s)"
        Label4.Text = "Outflow, O2 (m3/s)"
        Label5.Text = "Storage, S1 (m3)"
        Label6.Text = "Time change, delta T (hr)"
        Label7.Text = ""
        Button1.Text = "&Calculate"
    End Sub

    Private Sub Button1_Click(ByVal sender As
        System.Object, ByVal e As
        System.EventArgs) Handles Button1.Click
        Dim I1, I2, O1, O2, S1, S2, dT, dS As Double
        I1 = Val(TextBox1.Text)
        I2 = Val(TextBox2.Text)
        O1 = Val(TextBox3.Text)
        O2 = Val(TextBox4.Text)
        S1 = Val(TextBox5.Text)
        dT = Val(TextBox6.Text)
        dT *= 60 * 60              'convert to sec.

        dS = ((I1 + I2) / 2) - ((O1 + O2) / 2)
        dS *= dT
        S2 = S1 + dS
        Label7.Text = "Change in storage, dS = " +
            dS.ToString + " m3"
        Label7.Text += vbCrLf
        Label7.Text += "Storage volume after " +
            dT.ToString + ", S2 = " _
                    + S2.ToString + " m3"
    End Sub
End Class
```

2.5 Wind

Wind denotes air flowing nearly horizontally. Winds are mainly the result of horizontal differences in pressure. In absence of other factors tending to influence wind, it should be expected that its direction would be from areas of high pressure towards areas of low pressure and that its speed would vary with the pressure gradient.

Example 2.4

At a given site, a long-term wind speed record is available for measurements at heights of 10m and 15m above the ground. For certain calculations of evaporation the speed at 2m is required, so it is desired to extend the long-term record to the 2m level. For one set of data the speeds at 10 and 15 m were 7.68 and 8.11 m/s respectively (B.Sc., DU, 2013).

1. What is the value of the exponent relating the two speeds and elevations?
2. What speed would you predict for the 2 m level?

$$Q_o = \frac{\pi \times 20.5 \left(50^2 - 41.7^2 \right)}{Ln \dfrac{751.5}{\left(0.4/2 \right)}} = 5867 \quad \text{cubic meter per day}$$

Solution

1) Data given: $z_1 = 10$ m, $z_2 = 15$, $u_{10} = 7.68$ m/s, $u_{15} = 8.22$ m/s.
   Required: a, u_2

2) Use equation $\dfrac{u}{u_o} = \left(\dfrac{z}{z_o} \right)^a$ to determine the constant a:
   Where:
   u = Wind speed in the region at height zo m /s
   uo = Wind speed at least height (at anemometer) zo . m/s
   zo = Height of anemometer (minimum height), m
   a = An exponent varying with surface roughness and stability of the atmosphere, (usually ranging between 0.1 to 0.6 in the surface boundary layer).

$$\frac{7.68}{8.11} = \left(\frac{10}{15} \right)^a$$

52

Or: a*Log (10/15) = Log (7.68/8.11) which yields a = 0.134

3) Use equation

$$\frac{u_{10}}{u_2} = \left(\frac{z_{10}}{z_2}\right)^a \qquad\qquad \frac{7.68}{u_2} = \left(\frac{10}{2}\right)^{0.134}$$

Which yields: $u_2 = 6.19$ m/s

Example 2.4 Listing:

```
'***************************
'EXAMPLE 2.4: WIND
'***************************
Imports System.Math

Public Class Form1

    Private Sub Form1_Load(ByVal sender As
        System.Object, ByVal e As
        System.EventArgs) Handles MyBase.Load

        Me.Text = "Example 2.4: Wind"
        Me.MaximizeBox = False
        Me.FormBorderStyle =
        Windows.Forms.FormBorderStyle.FixedSingle

        Label1.Text = "Manometer min. height, z0 (m)"
        Label2.Text = "Wind speed at (z0), u0 (m/s)"
        Label3.Text = "Manometer max. height, z (m)"
        Label4.Text = "Wind speed at (z), u (m/s)"
        Label5.Text = "Predict speed at level of:"
        Label6.Text = ""
        Button1.Text = "&Calculate"
    End Sub

    Private Sub Button1_Click(ByVal sender As
        System.Object, ByVal e As
        System.EventArgs) Handles Button1.Click

        Dim u, u0, u2, z, z0, z2, a As Double
        z0 = Val(TextBox1.Text)
        u0 = Val(TextBox2.Text)
        z = Val(TextBox3.Text)
        u = Val(TextBox4.Text)
        z2 = Val(TextBox5.Text)

        a = (log10(u0 / u)) / (log10(z0 / z))
```

```
    u2 = u0 / ((z0 / z2) ^ a)

    Label6.Text = "Relation exponent (a) = " +
        FormatNumber(a, 3)
    Label6.Text += vbCrLf
    Label6.Text += "Predicted speed at " +
        z2.ToString + " = " _
                    + FormatNumber(u2, 2) + " m/s"
    End Sub
End Class
```

2.6 Theoretical Exercises

a) Write briefly about <u>three</u> of the following: (B.Sc., UAE, 1989).
   1) The **hydrologic cycle** and the principal influences that prevent its phases from repeating themselves in an identical fashion from year to year.
   2) Factors that influence **evaporation**. Explain briefly how each factor affects the rate and time of occurrence of evaporation.
   3) Types of **precipitation**. On what category would you place rain rainfall at El-Ain.
   4) Infiltration indexes and their importance.

b) Show how the following relationship can be used to find the **wind** speed from the height above the ground. (B.Sc., DU, 2011)

$$\frac{u}{u_o} = \left(\frac{z}{z_o}\right)^a$$

c) State Buys **Ballot** law (B.Sc., DU, 2013).

d) It is necessary to specify height above sea level when doing any measurement of **wind** due to surface friction factors and water surfaces through which wind is blowing. The relationship between wind speed and height can be found from the following relationship. Define the shown terms. (B.Sc., DU, 2012)

$$\frac{u}{u_o} = \left(\frac{z}{z_o}\right)^a$$

e) Define and explain: **Dew point, Ballton's** law, Orographic **precipitation** and an isohyetal map. (B.Sc., UAE, 1989, B.Sc., DU, 2012).

f) Define: **relative humidity, dew point** and **wind** (B.Sc., DU, 2012).

g) What is **hydrology**? What are the benefits of this science in practical life?

h) Indicate general factors affecting **climate**, and methods of measurement. Illustrate your answer with appropriate sketches.

i) Mention advantages of measuring **temperature**.

j) What is meant by "areas of low and high **pressure**"? How are they detected?.

k) Write a detailed report on each of the following: moisture, **relative humidity, water vapor pressure**, and **dew point**.

l) Outline marked differences between methods of measuring **humidity**.

m) What is the benefit of measuring solar **radiation**?

n) Explain ways of **cloud** formation.

o) Explain water vapor **condensation** in air. What are the benefits of the process?

p) What factors affect **evaporation**?

q) Write a detailed report on instruments for measuring **evaporation**.

r) How does both surface water and groundwater affect **evaporation**?

s) Mention different types of **precipitation** and their presence in practice.

t) How **precipitation** is measured? Compare its measurement devices.

u) What are elements of a hydrological **Meteorological station**?

v) What Is Precipitation? How is it measured? (B.Sc., UoD 2014)

w) Elaborate on types, forms and categories of precipitation in KSA. Comment on availability, continuity, accuracy, transparency and update of precipitation maps in the kingdom. Propose, to concerned department, suitable and appropriate remedial strategies. (B.Sc., UoD 2013)

x) The World Meteorological Organization, WMO, is a specialized agency of the United Nations. It is concerned with the state and behavior of the earth's atmosphere, its interaction with the oceans, the climate it produces and the resulting distribution of water resources. KSA is a member state with WMO, being the division hydrologist, show how you could utilize their resources for improving the meteorological systems at your division. (B.Sc., UoD 2013)

2.7 Problem solving in meteorological data
Pressure

a) A hill is of height of 4000 meters and of rising dry air temperature of 14.8 degrees Celsius. Condensation occurs at a height of 2000 meters. Calculate the **pressure** and saturation drop. Find air temperature when it reaches the top of the mountain and its temperature when it reaches top of hill and its temperature when it returns to the bottom.

Vapor pressure

b) An air mass is at a temperature of $18°$ C with relative humidity of 80 %. Using the following equation

$$h = 100 \; x \frac{e}{e_s}$$

find: (B.Sc., DU, 2012)
1. Saturation **vapor pressure**.
2. Actual vapor pressure in mm Hg, m bar, and Pa
3. The deficit in saturation
4. **Dew point.**

c) An air mass is at a temperature of 24.5 °C with relative humidity of 58%. Determine: (B.Sc., UAE, 1989).
1. Saturation **vapour pressure**.
2. Saturation deficit.
3. Actual vapour pressure (in mbar, mmHg, and Pa).
4. **Dew point**.
   (Ans. 23.05 mmHg, 13.37 mmHg, 9.68 mmHg, 15.7 °C)

d) A mass of air is at a temperature of 22°C and relative humidity of 81%. Find: saturated vapor pressure, real vapor pressure, saturation deficit, and dew point (Ans.   19.82,   16.05, 3.77mmHg, 18.6°C).

e) An air mass is at a temperature of $28°$ C with relative humidity of 70%. Find:
1. Saturation **vapor pressure**
2. Actual vapor pressure in mbar and mm Hg

3. Saturation deficit
4. **Dew point**
5. Wet-bulb temperature.

f) To a mass of air at a temperature of $20°$ C and relative humidity of 80%, find saturated vapor pressure, real vapor pressure, deficit in saturation, and dew point. What is the difference in your answer if the temperature changed to $25°$ C and relative humidity to 75%.

g) An air mass is at a temperature of $20°$ C with relative humidity of 75%. Using the following equation

$$h = 100 \; x \; \frac{e}{e_s}$$

find:

1. Saturation **vapor pressure**
2. Actual vapor pressure
3. The deficit in saturation
4. **Dew point**

    (Ans. 22.53 mm Hg, 16.8975 mm Hg, 5.63 mm Hg, 19.4 $°$ C).

Wind

h) Anemometers at 2.5 m and 40 m on a tower record wind speeds of 2 and 5 m/s respectively. Compute **wind speeds** at 5 and 30 m. (B.Sc., DU, 2012) (Ans. 2.51, 4.54 m/s)

i) At a given site, a long-term wind speed record is available for measurements at heights of 5 m and 10 m above the ground. For certain calculations of evaporation the speed at 2 m is required, so it is desired to extend the long-term record to the 2 m level. For one set of data the speeds at 5 and 10 m were 5.51 and 6.11 m/s respectively (B.Sc., DU, 2012).

1. What is the value of the exponent relating the two speeds and elevations?
2. What **wind speed** would you predict for the 2 m level?

    (Ans. 0.149, 4.8 m/s)

j) At a given site, a long-term wind speed record is available for measurements at heights of 5 m and 10 m above the ground. For certain calculations of evaporation the speed at 2 m is required, so it is desired to extend the long-term record to the 2 m level.

For one set of data the speeds at 5 and 10 m were 5.11 and 5.61 m/s respectively. (B.Sc., DU, 2011)

1. What is the value of the exponent relating the two speeds and elevations?
2. What **speed** would you predict for the 2 m level?

(Ans. 0.134677, 4.51 m/s)

k) Anemometers at 2m and 50m on a tower record wind speeds of 2 and 5 m/s respectively. Compute **wind speeds** at 5 and 30 m. (B.Sc., UAE, 1989). (2.6, 4.3 m/s)

l) Anemometers at 2.5 m and 40 m on a tower record wind speeds of 2 and 5 m/s respectively. Compute **wind speeds** at 5 and 30 m. (ans. 2.51, 4.54 m/s).

m) Anemometers at 2m and 50m on a tower record wind speeds of 2 and 5 m/s respectively. Compute **wind speeds** at 5 and 30 m. (ans. 2.6, 4.6 m/s).

n) Theoretical Wind speed is measured for two heights 3 and 4 meters and the following values were found 2.5 and 3 m/s, respectively. Find wind speed for a height of two meters.

o) Measured wind speed in a monitoring station for two heights 1 and 2 meters has shown the values of 1.6 and 2.2 m/s, respectively. Find **wind speed** at a height of two meters and ten meters.

Water budget

p) At a water elevation of 6391 ft, Kenyir Lake has a volume of 2, 939, 000 ac-ft, and a surface area, A of 48, 100 ac. Annual inputs to the lake include 8.0 in. of direct precipitation, runoff from gauged streams of 150, 000 ac-ft per year, and ungauged runoff and groundwater inflow of 37, 000 ac-ft per year. Evaporation is 45 inch per year. Make a **water budget** showing inputs, in ac-ft per year. (*Note : 1 ac = 43, 560 ft^2, 1 cm = 0.394 in, 1 in = 0.083 ft)   (Ans. 39, 284.90 ac-ft/year)

Chapter Three

Precipitation

3.1 Precipitation

Precipitation denotes all kinds of rainfall on surface of earth from vapor in the atmosphere (most of humidity at a height of 8 km from the Earth's surface).

Table 3.1 Forms of precipitation [1,2, 10]

Form of precipitation	Rate of fall (mm/hr)
Light	2.5
Moderate	2.8 – 7.6
Heavy	> 7.6

3.2 Interpolation of rainfall records

Sometimes records may be lost from the measuring or monitoring station for a specific day or several days because of the absence of station operator (observer) or because of instrumental failure or malfunction or damage in the recording devices, for any other reason. In order not to lose information, it is best to use an appropriate way to estimate the amount of rain in these days in calculating monthly and annual totals. Procedure for these estimates are based depending on simultaneous records for three stations close to and as evenly spaced around the station with missing records as possible. This station should be equi-distant from the three stations and the following conditions should be achieved:

1) If the normal annual precipitation at each of the these stations is within ten percent of that of the station with missing records, a simple arithmetic average of the precipitation at

the three stations is used for estimating missing record of the station.

2) If the normal annual precipitation at any one of the three stations differs from that of the station with missing records by more than ten percent, the normal ratio method is used [15].

Example 3.1

1) Rain gauge X was out of operation for a month during which there was a storm. The rainfall amounts at three adjacent stations A, B, and C were 37, 42 and 49 mm. The average annual precipitation amounts for the gauges are X = 694, A = 726, B = 752 and C = 760 mm. Using the Arithmetic method, estimate the amount of rainfall for gauge X.

Stations	Amounts of precipitation (mm)	Normal Annual Precipitation (mm)
A	37	726
B	42	752
C	49	760
D	?	694

Solution

If

$N_x = 694$

Then 10% from 694 = (10/100) * 694 = 69.4mm

Therefore, precipitation allowed

= (694 – 69.4) ~ (694 + 69.4) mm

= 624.6 mm ~ 763.4 mm

Since all annual precipitations (726, 752 and 760) mm   are within the ranges, **Arithmetic Method** can be applied :

$$P_x = \frac{1}{3}[37 + 42 + 49] = 42.7 \, mm$$

Program 3.1 Listing:

```
'********************************
'EXAMPLE 3.1: Rainfall estimate
'********************************
Public Class Form1

    Private Sub Form1_Load(ByVal sender As
        System.Object, ByVal e As
        System.EventArgs) Handles MyBase.Load

        Me.Text = "Example 3.1"
        Me.FormBorderStyle =
        Windows.Forms.FormBorderStyle.FixedSingle
        Me.MaximizeBox = False

        Label1.Text = "Amount of precipitation (mm)"
        Label2.Text = "Normal annual precip. (mm)"
        Label3.Text = "(A)"
        Label4.Text = "(B)"
        Label5.Text = "(C)"
        Label6.Text = "(X)"
        Label7.Text = "Select method:"
        Label8.Text = ""
        TextBox8.Enabled = False
        Button1.Text = "&Estimate rainfall"
        ComboBox1.Items.Clear()
        ComboBox1.Items.Add("Arithmetic method")
        ComboBox1.Items.Add("Example 2.2 Equation")
    End Sub

    Private Sub Button1_Click(ByVal sender As
        System.Object, ByVal e As
        System.EventArgs) Handles Button1.Click

        Dim Pa, Pb, Pc, Px As Double
        Dim Na, Nb, Nc, Nx As Double

        Pa = Val(TextBox1.Text)
        Pb = Val(TextBox4.Text)
        Pc = Val(TextBox6.Text)
        'Px = Val(TextBox8.Text)
        Na = Val(TextBox2.Text)
        Nb = Val(TextBox3.Text)
        Nc = Val(TextBox5.Text)
        Nx = Val(TextBox7.Text)

        Select Case ComboBox1.SelectedIndex
            Case 0'calculate using arithmetic equation
```

```
        Dim min, max As Double
        min = Nx - (Nx / 10)'(average - 10%)
        max = Nx + (Nx / 10)'(average + 10%)
        If (Na >= min And Na <= max) And _
           (Nb >= min And Nb <= max) And _
           (Nb >= min And Nb <= max) Then
           Px = (Pa + Pb + Pc) / 3
           Label8.Text = "All precipitations
within ranges " _
+ FormatNumber(min, 2) + "-"
+ FormatNumber(max, 2)
           Label8.Text += vbCrLf
           Label8.Text += "Using arithmetic
method:" + vbCrLf
           Label8.Text += "Px = " +
Px.ToString + " mm"
         Else
           Label8.Text = "Precipitations
are not within ranges " _
+ FormatNumber(min, 2) + "-"
+ FormatNumber(max, 2)
           Label8.Text += vbCrLf
           Label8.Text += "Cannot use
arithmetic method"
         End If
      Case 1'calculate using Example 2.2 Equation
         Px = ((Nx * Pa / Na) + (Nx * Pb / Nb)
+ (Nx * Pc / Nc)) / 3
         Label8.Text = "Px = " + Px.ToString
+ " mm"
    End Select
  End Sub
End Class
```

Example 3.2 (see Program 3.1 Listing)

One of four monthly-read rain gauges on a catchment area develops a fault in a month when the other three gauges record 48, 58 and 69 mm respectively. If the average annual precipitation amounts of these three gauges are 741, 769 and 855 mm respectively and of the broken gauge 707 mm, estimate the missing monthly precipitation at the latter (B.Sc., DU, 2013).

$$P_x = \frac{\dfrac{N_x P_a}{N_a} + \dfrac{N_x P_b}{N_b} + \dfrac{N_x P_c}{N_c}}{3}$$

Solution

1. Data: $P_a = 48$ mm, $P_b = 58$ mm, $P_c = 69$ mm, $N_a = 741$ mm, $N_b = 769$ mm, $N_c = 855$ mm, $N_x = 707$ mm.
2. Find the value of rainfall during the storm at station (a) by using the following equation:

$$P_x = \frac{\dfrac{N_x P_a}{N_a} + \dfrac{N_x P_b}{N_b} + \dfrac{N_x P_c}{N_c}}{3} = \frac{\dfrac{707 \times 48}{741} + \dfrac{707 \times 58}{769} + \dfrac{707 \times 69}{855}}{3}$$

= 52 mm

Example 3.3 (see Program 3.1 Listing)

The records of precipitation of hydraulic monitoring stations (x) in a rainy day are missing. The data indicate that the estimates of rainfall at three stations (b), (c) and (d) adjacent to the station (x) are equal to: 80, 70 and 60 mm, respectively. If the average annual rainfall at stations (a) and (b) and (c) and (d) is: 650, 240, 320 and 140 mm, respectively, find the value of rainfall during the rain storm in station (x) (B.Sc., DU, 2012) .

Solution

- Data: Pb = 80 mm, Pc = 70 mm, Pd = 60 mm, Nx = 650 mm, Nb = 240 mm, Nc = 320 mm, Nd = 140 mm.
- Find the value of rainfall during the storm at station (a) by using the following equation: Px = 212.5 mm

3.3 Methods used to find the height of rainfall

Methods used to find the height of rainfall Include: Arithmetic mean method, Thiessen polygon method and isohyets method.

Example 3.4

Using the data given below, estimate the average precipitation using Thiessen Polygon method.

Stations	Area (km^2)	Precipitation (mm)	Area x Precipitation (km^2·mm)
A	72	90	6 480
B	34	110	3 740
C	76	105	7 980
D	40	150	6 000
E	76	160	12 160
F	92	140	12 880
G	46	130	5 980
H	40	135	5 400
I	86	95	8 170
J	6	70	420
TOTAL	568	1 185	69 210

Solution

$$Average\ Precipitation = \frac{\sum Area\ for\ each\ Station\ x\ Precipitation}{\sum Total\ Polygon\ Area}$$

$$Average\ Precipitation = \frac{69210}{568}$$

$$Average\ Precipitation = 121.8\ mm$$

Program 3.4 Listing:

```
Public Class Form1

    Private Sub Form1_Load(ByVal sender As
        System.Object, ByVal e As
        System.EventArgs) Handles MyBase.Load

        Me.Text = "Example 3.4: Thiessen polygon"
        Me.MaximizeBox = False
        Me.FormBorderStyle =
        Windows.Forms.FormBorderStyle.FixedSingle

        Label1.Text = "Select option:"
        Label4.Text = ""
        ComboBox1.Items.Clear()
        ComboBox1.Items.Add("Find avg. precipitation")
        ComboBox1.Items.Add("Find missing recording")
        ComboBox1.SelectedIndex = 0
        Button1.Text = "&Calculate"
    End Sub
```

```vb
Private Sub ComboBox1_SelectedIndexChanged(ByVal
    sender As System.Object, ByVal e As
    System.EventArgs) Handles
    ComboBox1.SelectedIndexChanged

    'adjust user interface according to selection
    Select Case ComboBox1.SelectedIndex
        Case 0
            Label2.Text = "Enter station data
    (Area, Precipitation):"
            Label3.Text = ""
            TextBox1.Visible = False
            DataGridView1.Columns.Clear()
            DataGridView1.Columns.Add("areaCol",
        "Area (km2)")
            DataGridView1.Columns.Add("preCol",
        "Precipitation (mm)")
            DataGridView1.Columns.Add("apCol",
        "Area*Precip.")
            DataGridView1.Columns(2).ReadOnly = True
        Case 1
            Label2.Text = "Enter station data
        (Recording, Thiessen),
        leave missing recording
        empty:"
            Label3.Text = "Mean precipitation ="
            TextBox1.Visible = True
            DataGridView1.Columns.Clear()
            DataGridView1.Columns.Add("rainCol",
        "Rainfall (mm)")
            DataGridView1.Columns.Add("thiessenCol",
        "Thiessen polygon (km2)")
            DataGridView1.Columns.Add("rtCol",
        "(1)*(2)")
            DataGridView1.Columns(2).ReadOnly = True
    End Select
    Label4.Text = ""
End Sub

Private Sub Button1_Click(ByVal sender As
    System.Object, ByVal e As
    System.EventArgs) Handles Button1.Click

    Dim area, ap As Double
    Dim tmp, tmp2, tmp3 As Double
    Select Case ComboBox1.SelectedIndex
        Case 0 '*****calculate avg. precipitation
            area = 0 : ap = 0
            If DataGridView1.RowCount <= 1 Then
```

```
         MsgBox("Enter at least one
reading!", vbOKOnly, "Error")
         Exit Sub
      End If

      For i = 0 To DataGridView1.RowCount - 2
         tmp =
   Val(DataGridView1.Rows(i).
      Cells(0).Value) * _
   Val(DataGridView1.Rows(i).
      Cells(1).Value)
         area +=
   Val(DataGridView1.Rows(i).
      Cells(0).Value)
         ap += tmp
         DataGridView1.Rows(i).Cells(2).
      Value = tmp
      Next
'divide Area*Precip. by Total Area
      ap /= area
      Label4.Text = "Average
precipitation = " + _
         FormatNumber(ap, 2) + " mm"
         '*************************************
   Case 1 '*****find missing recording
      area = 0 : ap = 0
      If DataGridView1.RowCount <= 1 Then
         MsgBox("Enter at least one
   reading!", vbOKOnly, "Error")
         Exit Sub
      End If

      For i = 0 To DataGridView1.RowCount - 2
         tmp =
   Val(DataGridView1.Rows(i).
      Cells(0).Value) * _
   Val(DataGridView1.Rows(i).
      Cells(1).Value)
         area +=
   Val(DataGridView1.Rows(i).
      Cells(1).Value)
         ap += tmp
         DataGridView1.Rows(i).Cells(2).
   Value = tmp
         'if the product value is zero,
   'then this is the missing
   'reading, so save it's area in tmp3
         'to use it in calculation later.
         If tmp = 0 Then tmp3 =
   Val(DataGridView1.Rows(i).
```

66

```
                    Cells(1).Value)
             Next
       'value of mean precip.
             tmp2 = Val(TextBox1.Text)
             tmp2 *= area
             tmp2 -= ap
             tmp2 /= tmp3
             Label4.Text = "Missing recording = "+
                  FormatNumber(tmp2, 2) + " mm"
       End Select
    End Sub
End Class
```

Example 3.5 (see Program 3.4 Listing)

The precipitation on a catchment in Dubai of area 95 km^2 is sampled in table (1). Determine the precipitation recorded by station number (7) if the mean precipitation, as computed by Thiessen method, amounts to 98mm. (B.Sc., UAE, 1989).

Rain gauge	Recorded rainfall, Feb. 1968 (mm)	Thiessen polygon on area (km^2)
1	84	4.0
2	90	4.0
3	120	10.0
4	86	5.1
5	87	15.1
6	76	30.6
7	X	6.2
8	131	20.0

Solution

a)

Rain gauge	Recorded rainfall, Feb. 1968 (mm)	Thiessen polygon on area (km^2)	Product
1	84	4.0	336
2	90	4.0	360
3	120	10.0	1200
4	86	5.1	438.6

5	87	15.1	1313.7
6	76	30.6	2325.6
7	X	6.2	6.2x
8	131	20.0	2620

$$\sum A = 95$$

$$\sum AP = 8593.9 + 8.2\,x$$

$$98 = \frac{8593.9 + 6.2\,x}{95}$$

$$x = \frac{716.1}{6.2} = 115.5\ mm$$

Example 3.6

Find area precipitation by isohyetal method for a certain catchment area given the following data (B.Sc., DU, 2012).

Isohyet, in	Area enclosed within basin boundary, sq. mile
6.8	
6	20
5	97
4	213
3	410
2	602
1.5	633

Solution

Isohyet, in Col. (1)	Area enclosed within basin boundary, sq. mile Col. (2)	Net area, sq. mile Col. (3) (b_n-b_{n-1})	Average precipitation Col. (4) $(a1+a2)/2$	Precipitation volume Col. (3)*Col. (4)
6.8				
6	20	20	$(6.8+6)/2 =$ 6.4	128
5	97	97-20 = 77	5.5	423.5

4	213	213-97 = 116	4.5	522
3	410	410-213 = 197	3.5	689.5
2	602	602-410 =192	2.5	480
1.5	633	633-602 = 31	1.75	54.25
	Total area = 633			Total volume = 2297.25

P = 2297.25/633 = 3.63 in

Program 3.6 Listing:

```
'***********************************
'EXAMPLE 3.6: Isohyetal method
'***********************************
Public Class Form1

    Private Sub Form1_Load(ByVal sender As
        System.Object, ByVal e As
        System.EventArgs) Handles MyBase.Load

        Me.Text = "Example 3.6: Isohyetal method"
        Me.MaximizeBox = False
        Me.FormBorderStyle =
    Windows.Forms.FormBorderStyle.FixedSingle

        Label1.Text = "Select option:"
        Label4.Text = ""
        ComboBox1.Items.Clear()
        ComboBox1.Items.Add("Find area precipitation")
        ComboBox1.Items.Add("Estimate mean basin
            precipitation")
        ComboBox1.SelectedIndex = 0
        Button1.Text = "&Calculate"
    End Sub

    Private Sub ComboBox1_SelectedIndexChanged(ByVal
        sender As System.Object, ByVal
        e As System.EventArgs) Handles
        ComboBox1.SelectedIndexChanged

        'adjust user interface according to selection
        Select Case ComboBox1.SelectedIndex
```

```
        Case 0
            Label2.Text = "Enter data for
        isohyets and area enclosed
        within basin boundary:"
            Label3.Text = ""
            TextBox1.Visible = False
            DataGridView1.Columns.Clear()
            DataGridView1.Columns.Add("isoCol",
        "Isohyet (in)")
            DataGridView1.Columns.Add("areaCol",
        "Area within basin boundary
        (sq. mi)")
            DataGridView1.Columns.Add("netAreaCol",
        "Net Area")
            DataGridView1.Columns.Add(
        "avgPrecipCol", "Avg. precip.")
            DataGridView1.Columns.Add(
        "precipVolCol", "Precip. volume")
            DataGridView1.Columns(2).ReadOnly = True
            DataGridView1.Columns(3).ReadOnly = True
            DataGridView1.Columns(4).ReadOnly = True
        Case 1
            Label2.Text = "Enter data for
        isohyets, area within basin
        boundary, and mean precipitation:"
            Label3.Text = ""
            TextBox1.Visible = False
            DataGridView1.Columns.Clear()
            DataGridView1.Columns.Add("isoCol",
        "Isohyet (in)")
            DataGridView1.Columns.Add("areaCol",
        "Area within basin boundary (ha)")
            DataGridView1.Columns.Add(
        "meanPrecipCol", "Mean precip.
        within isohyet")
            DataGridView1.Columns.Add("netAreaCol",
        "Net Area")
            DataGridView1.Columns(3).ReadOnly = True
    End Select
    Label4.Text = ""
End Sub

Private Sub Button1_Click(ByVal sender As
    System.Object, ByVal e As
    System.EventArgs) Handles Button1.Click

    If DataGridView1.RowCount <= 1 Then
        MsgBox("Enter at least one reading!",
        vbOKOnly, "Error")
        Exit Sub
```

```vb
End If
Dim area, netArea, avg, vol As Double
Dim tmp, tmp2, tmp3 As Double
Select Case ComboBox1.SelectedIndex
    Case 0 '*****calculate area precipitation
        area = 0 : netArea = 0 : avg = 0 :
        vol = 0
        For i = 1 To DataGridView1.RowCount - 2
            tmp =
    Val(DataGridView1.Rows(i).
    Cells(1).Value) - _
            Val(DataGridView1.Rows(i -
    1).Cells(1).Value)
            area += tmp
            DataGridView1.Rows(i).Cells(2).
    Value = tmp
            tmp2 =
    (Val(DataGridView1.Rows(i).
        Cells(0).Value) + _
                Val(DataGridView1.Rows(i -
        1).Cells(0).Value))/2
            avg += tmp2
            DataGridView1.Rows(i).Cells(3).
    Value = tmp2
            tmp3 = tmp * tmp2
            vol += tmp3
            DataGridView1.Rows(i).Cells(4).
    Value = tmp3
        Next
    'divide Total Precip. by Total Area
        vol /= area
        Label4.Text = "Area precipitation = "+
        FormatNumber(vol, 2) + " in."
        '***************************************
        '***************************************

    Case 1 '*****estimate mean basin
    '*****precipitation
        area = 0 : netArea = 0 : avg = 0 :
        vol = 0
        For i = 0 To DataGridView1.RowCount - 2
            If i = 0 Then
                tmp =
    Val(DataGridView1.Rows(i).
    Cells(1).Value)
            Else
                tmp =
    Val(DataGridView1.Rows(i).
        Cells(1).Value) - _
                Val(DataGridView1.Rows(i -
        1).Cells(1).Value)
```

71

```
                    End If
                    area += tmp
                    DataGridView1.Rows(i).Cells(3).
            Value = tmp
                    tmp2 = tmp *
            Val(DataGridView1.Rows(i).
            Cells(2).Value)
                    vol += tmp2
                Next
            'divide Total Precip. by Total Area
                vol /= area
                Label4.Text = "Mean precipitation = "+
                    FormatNumber(vol, 2) + " in."
        End Select
    End Sub
End Class
```

Example 3.7 (see Program 3.6 Listing)

The following table shows rainfall observations taken for a certain area. Estimate the mean basin precipitation (B.Sc., DU, 2012) .

Table: Rainfall observations.

Isohyets (in)	Total area enclosed within basin boundary (ha)	Estimated mean precipitation between isohyets
> 6	50	6.3
> 5	120	5.5
> 4	250	4.7
> 3	450	3.6
> 2	780	2.7
> 1	999	1.4
< 1	1020	0.8

Solution

Determine corresponding net area as in table below:

Isohyets (in)	Total area enclosed within basin boundary (ha)	Corresponding net area	Estimated mean precipitation between isohyets
> 6	50	50	6.3

> 5	120	70	5.5
> 4	250	130	4.7
> 3	450	200	3.6
> 2	780	330	2.7
> 1	999	219	1.4
< 1	1020	21	0.8

$$n_i = \frac{\sum_{i=1}^{n} A_i \times \frac{P_{i-1} + P_i}{2}}{\sum_{i=1}^{n} A_i}$$

= 50x6.3 + 70x5.6 + 130x4.7 + 200x3.6 + 330x3.7 + 319x1.4
+ 31x0.8 + 50 + 70 + 130 + 200 + 330 + 219 + 21 = 11.06"

Example 3.8 (see Program 3.6 Listing)

The following rainfall observations were taken for a certain area: (B.Sc., UAE, 1989).

Isohyets (in)	Total area enclosed within basin boundary (ha)	Estimated mean precipitation between isohyets
> 6	40	6.2
> 5	110	5.5
> 4	236	4.5
> 3	430	3.5
> 2	772	2.5
> 1	990	1.5
< 1	1010	0.9

Estimate the mean basin precipitation

Solution

Determine corresponding net area as presented in the following table:

Isohyets (in)	Total area enclosed within basin boundary (ha)	Corresponding net area	Estimated mean precipitation between isohyets
> 6	40	40	6.2
> 5	110	70	5.5
> 4	236	126	4.5
> 3	430	194	3.5
> 2	772	342	2.5
> 1	990	218	1.5
< 1	1010	20	0.9

$$n_i = \frac{\sum\limits_{i=1}^{n} A_i \times \dfrac{P_{i-1}+P_i}{2}}{\sum\limits_{i=1}^{n} A_i}$$

$$= \frac{40 \, x \, 6.2 + 70 \, x \, 5.5 + 126 \, x \, 4.5 + 194 \, x \, 3.5 + 342 \, x \, 2.5 + 218 \, x \, 1.5 + 20 \, x \, 0.9}{40 + 70 + 126 + 194 + 342 + 218 + 20}$$

$$= \frac{3079}{1010} = 3.05'' \approx 3''$$

Example 3.9 (see Program 3.6 Listing)

Use the isohyetal method to determine the average precipitation depth within the basin for the storm.

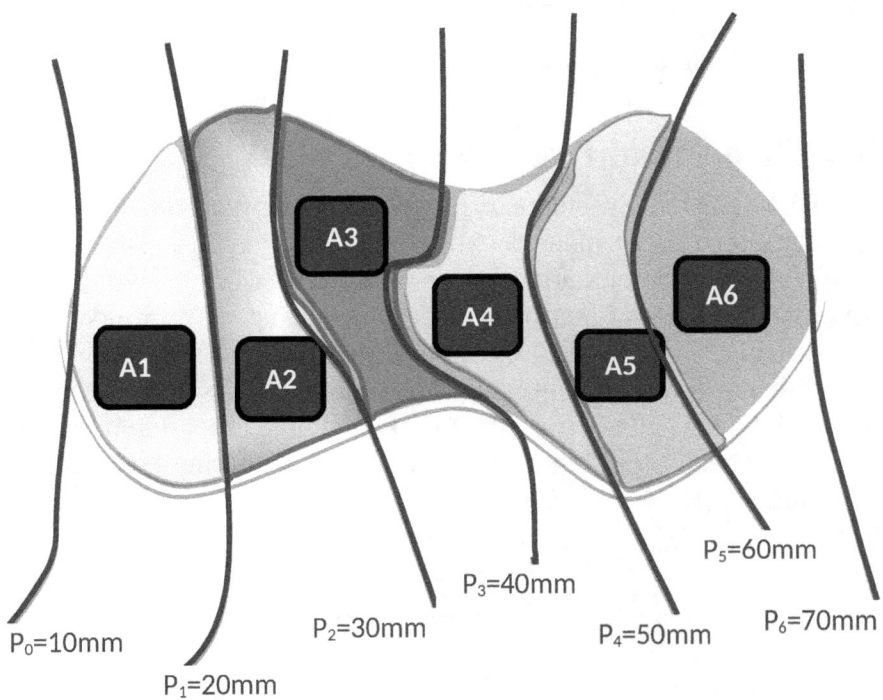

$P_5=60mm$

$P_3=40mm$

$P_2=30mm$ $P_4=50mm$ $P_6=70mm$

$P_0=10mm$

$P_1=20mm$

Solution

Isohyetal interval	Average Precipitation (mm)	Area $(km^{2.})$	Area x Precipitation $(km^2 \cdot mm)$
< 10	10	0	0
10 – 20	15	84	1 260
20 – 30	25	75	1 875
30 – 40	35	68	2 380
40 – 50	45	60	2 700
50 – 60	55	55	3 025
60 – 70	65	86	5 590
TOTAL		428	16 830

$$Average\ Precipitation = \frac{\sum Area \times Precipitation}{\sum Area}$$

$$Average\ Precipitation = \frac{16830}{428}$$

$Average\ Precipitation = 39.3\ mm$

3.4 Theoretical Exercises

a) Distinguish between the different forms of **precipitation**.

b) Outline sources of error when recording readings and record-keeping by precipitation **gauge** (B.Sc., DU, 2011)

c) List five natural and man-made factors that may influence the distribution of **rainfall** in a given area. (B.Sc., UAE, 1989).

d) Indicate how the following equation illustrates the method of weighting, by the ratio of the normal annual precipitation values to estimate missing records for a neighboring **monitoring station**. (B.Sc., DU, 2011)

$$P_x = \frac{\frac{N_x P_a}{N_a} + \frac{N_x P_b}{N_b} + \frac{N_x P_c}{N_c}}{3}$$

e) Define the terms used in the equation that illustrates the method of weighting to estimate **missing records** for a neighboring monitoring station (B.Sc., DU, 2012).

$$P_x = \frac{\frac{N_x P_a}{N_a} + \frac{N_x P_b}{N_b} + \frac{N_x P_c}{N_c}}{3}$$

f) What are the expected sources of error when recording **precipitation** readings in hydrological meteorological stations?

g) Explain methods of **rain** calculation? Indicate advantages and shortcomings of each method.

h) How do you assess **rainfall** data lost in a station as compared to data allocated to the neighboring stations?

i) What is the relationship between **intensity, duration and frequency of rainfall**?

3.5 Problem solving in precipitation

Arithmetic average rainfall

1) Find **average rainfall** for the rainfall data observed in the hydrological measuring stations described in the following table

Station number	Amount of rainfall, mm
1	250
2	380
3	490
4	310

2) Average rainfall in a specific region is equal to 310 mm according to the data shown in the table below for estimated rainfall in four hydrological monitoring stations: What is the **average rainfall** in station number (4) using arithmetic average method for estimating rainfall?

Station number	Amount of rain, mm
1	300
2	400
3	250
4	?

Thiesen polygon

3) Find average precipitation in watershed according to the information and data that are described in the following table using average arithmetic method and **Thiesen polygon**. (Ans. 100.3,).

Station	Rain gauge, cm	Area, %
1	62	8
2	74	11
3	87	20
4	112	18
5	118	33
6	107	10

4) List five natural and man-made factors that may influence the distribution of rainfall in a given area. The precipitation on a catchment in Dubai of area 95 km^2 is sampled in table (1). Determine the precipitation recorded by station number (7) if

the mean precipitation, as computed by **Thiessen method**, amounts to 98mm. (ans. 115.5 mm).

Rain gauge	Recorded rainfall, Feb. 1968 (mm)	Thiessen polygon on area (km^2)
1	84	4.0
2	90	4.0
3	120	10.0
4	86	5.1
5	87	15.1
6	76	30.6
7	X	6.2
8	131	20.0

5) Find precipitation by **Thiessen method** for the observed precipitation records shown in table (4) (B.Sc., DU, 2013).

$$P_{mean} = \frac{\sum_{i=1}^{n} A_i P_i}{\sum_{i=1}^{n} A_i}$$

Table: Precipitation records.

Observed precipitation, in	Area of corresponding polygon within basin boundary, sq. mile
1.59	12
2.36	112
2.84	109
3.61	122
2.46	19
3.9	84
6.11	94
5.61	68

Solution

Observed precipitation, in (1)	Area of corresponding polygon within basin boundary, sq. mile (2)	Percent total area (3)	Weighted precipitation, in Col. (1) x Col. (3)
1.59	12	1.9	0.03

2.36	112	18.1	0.43
2.84	109	17.6	0.5
3.61	122	19.7	0.71
2.46	19	3.1	0.08
3.9	84	13.5	0.53
6.11	94	15.1	0.93
5.61	68	11	0.62
	Total = 620	100	3.83

Average P = 3.83 in

Method of isohyets

6) Calculate the average rainfall of the data set in the following table using **isohyet method** rainfall.

Isohyet, cm	Area between isohyets, km^2
5	10
10	44
15	50
20	60
25	10
30	24
35	

7) Find average rainfall for the following data using **isohytal method**,

Isohyet, cm	Area between isohyets, km^2
5	15
6	25
7	95
8	105
9	55
10	25
11	

8) Find average precipitation in catchment area according to the information and data that are described in the following table using the **method of isohyets**.

79

Isohyets, inch	total area bounded by boundaries of the basin (ha)	Estimated average rain between isohyets, inch
6 <	40	6.5
5 <	110	5.5
4 <	236	4.5
3 <	430	3.5
2 <	772	2.5
1 <	990	1.5
< 1	1010	0.9

9) Table shows rainfall observations taken for a certain area. Estimate the mean basin precipitation. (ans. 11.06").

Isohyets (in)	Total area enclosed within basin boundary (ha)	Estimated mean precipitation between **isohyets**
> 6	50	6.3
> 5	120	5.5
> 4	250	4.7
> 3	450	3.6
> 2	780	2.7
> 1	999	1.4
< 1	1020	0.8

10) The following rainfall observations were taken for a certain area:

Isohyets (in)	Total area enclosed within basin boundary (ha)	Estimated mean precipitation between **isohyets**
> 6	40	6.2
> 5	110	5.5
> 4	236	4.5
> 3	430	3.5
> 2	772	2.5
> 1	990	1.5
< 1	1010	0.9

Estimate the mean basin precipitation (ans. 3').

Rainfall intensity

11) Find average rainfall intensity over an area of 6 square kilometers during a one-hour storm of frequency of once every ten years (assuming point rain of 30 mm, and inverse gamma 5.6) (Ans. 26 mm).

12) What is the average rainfall intensity over an area of 5 square kilometers during a 60 minute storm with a frequency of once in 10 years (take p = 25 mm).

13) From the precipitation data given, estimate cumulative rainfall and rainfall **intensity**.

Time (min)	0	10	20	30	40	50	60	70	80	90
Rainfall (cm)	0	0.18	0.21	0.26	0.32	0.37	0.43	0.64	1.14	3.18

Time (min)	100	110	120	130	140	150	160	170	180
Rainfall (cm)	1.65	0.81	0.52	0.26	0.42	0.36	0.28	0.19	0.17

API

14) The value of the **API** for a certain station reached 42 mm in first of August, and an amount of 46 mm of rain fell on the fifth of August, also rain of 28 mm fell in the seventh of august and 34 mm on the eighth day of August. Find value of the API guide on 12th August noting that k is equal to 0.92. Calculate the amount of the index assuming no rain in the same period. Draw curve of change of index with time. (Ans. 85.3, 1.7 mm).

15) The value of the **API** for a station reached 55 mm in the first of August, an amount of 57 mm of rain fell on the sixth of August, 31 mm of rain fell in the eighth of August and 20 mm fell in the ninth day of August. Find value of API index for the 15th day of August, noting that k is equal to 0.92. Calculate the amount of index assuming no rain during the same period. Draw curve of change of index with time.

16) The antecedent precipitation index (API) is given by: $I_e = I_o k^t$, Define the parameters shown in the equation. The API for a station was 60 mm on the first of December, 64 mm rain fell on the seventh of December, 40 mm rain fell on the ninth of December, and 30 mm rain fell on the eleventh of December. Compute the API for 20th December if k = 0.85. Determine the **API** if no rain fell. (ans. 24.1, 2.7 mm).

81

Missing record

17) A precipitation station (X) was inoperative for some time during which a storm occurred. The storm totals at three stations (A), (B) and (C) surrounding X, were respectively 6.60, 4.80 and 3.30 cm. The normal annual precipitation amounts at stations X, A, B and C are respectively 65.6, 72.6, 51.8 and 38.2 cm. Estimate storm precipitation for station (X).

18) Rainfall data taken from the records of hydrological monitoring stations (a), (b), and (c), and (d) gave values of 17, 18, 24 and 18 mm respectively. Thiesen polygon method was chosen to calculate the average rainfall. The area of each of the polygons surrounding each station is as follows: 24, 18, 32 and 35 square kilometers for stations (a), (b), and (c), (d), respectively. Find average amount of rainfall in the region.

19) Rain record is missed from a hydraulic station (m) in a windy day. However, the estimates of rainfall at three stations (m1), (m2) and (m3) surrounding station (m) is equal to 25, 50 and 35 mm respectively. Note that the average annual precipitation at the stations (m), (m1), (m 2) and (m3) equals 500 and 700 and 590 and 440 mm, respectively. Find the value of rainfall during the storm at station (m).

20) Four hydrological monitoring stations are of similar conditions registered the data shown in the following table. Estimate missing recod.

Station	1	2	3	4
Mean annual precipitation, mm	350	1200	1100	1000
Present precipitation, mm	750	?	760	810

21) Outline sources of error when recording readings and record-keeping by precipitation gauge. Indicate how the following equation illustrates the method of weighting, by the ratio of the normal annual precipitation values to estimate missing records for a neighboring monitoring station.

$$P_x = \frac{\dfrac{N_x P_a}{N_a} + \dfrac{N_x P_b}{N_b} + \dfrac{N_x P_c}{N_c}}{3}$$

The records of precipitation of hydraulic monitoring stations (x) in a rainy day are missing. The data indicate that the estimates of rainfall at three stations (b), (c) and (d) adjacent to the station (x) are equal to: 70, 608 and 50 mm, respectively. If the average annual rainfall at stations (a) and (b) and (c) and (d) is: 600, 250, 510 and 120 mm, respectively, find the value of rainfall during the rain storm in station (x). (ans. 162.9 mm).

22) The average annual precipitation amounts for the gauges A, B, C and D are 1120, 935, 1200 and 978 mm. In year 1975, station D was out of operation. Station A, B, and C recorded rainfall amounts of 107, 89 and 122 mm respectively. Estimate the amount of precipitation for station D in year 1975.

Stations	Normal Annual Precipitation (mm)	Amounts of Precipitation Year 1975 (mm)
A	1120	107
B	935	89
C	1200	122
D	978	X

(Ans. 880.2 mm ~ 1075.8 mm)

Since the average annual precipitations   amounts for the gauges   A and C exceeded 1075.8 mm, therefore **Normal Ratio Method** is used :-

$$P_x = \frac{978}{3}\left[\frac{107}{1120}+\frac{89}{935}+\frac{122}{1200}\right] = 95.3\, mm$$

23) The records of precipitation of hydraulic monitoring stations (x) in a rainy day are missing. The data indicate that the estimates of rainfall at three stations (b), (c) and (d) adjacent to the station (x) are equal to: 70, 608 and 50 mm, respectively. If the average annual rainfall at stations (a) and (b) and (c) and (d) is: 600, 250, 510 and 120 mm, respectively, find the value of rainfall during the rain storm in station (x). (B.Sc., DU, 2011) **(Ans. 162.9 mm)**

24) The precipitation amounts for the months of October, November, and December are missing from the record for one gauging station in a drainage basin. This station belongs to a network of four in that catchment area. For those three months,

the other three stations recorded the following: (B.Sc., UoD 2014)

	Station		
	1	2	3
October	67	77	87
November	59	62	57
December	57	52	67

Estimate the missing precipitation values if the long-term annual average precipitation at the four stations is:

	Station			
	1	2	3	4
October	72	77	82	79
November	62	67	77	72
December	57	59	72	67

Chapter Four

Evaporation and transpiration

4.1 Evaporation

Evaporation is the process by which water is changed from the liquid or solid state into the gaseous state through the transfer of heat energy [15]. Evapotranspiration: the process by which water is evaporated from wet surfaces and transpired by plants [15].

Methods of estimating evaporation or transpiration include: Evaporation pans, empirical formulas, methods of water budget, methods of mass transfer and methods of energy budget [13].

Example 4.1

The daily potential evapotranspiration from a field crop at latitude 20° N in November has been predicted through use of the nomogram for the solution of Penman's equation to be equal to 2.8 mm. Estimate the mean wind speed at height of 2m, u_2, m/s. Governing conditions are as outlined in the following table (B.Sc., DU, 2013).

Table: Evaporation data.

Item	Value
Mean air temperature, T, °C	22
Sky cover	70 % cloud
mean humidity, h, %	60
Ratio of potential evapotranspiration to potential evaporation (E_o)	0.8

$$E_T = E_1\left(t, \frac{n}{D}\right) + E_2\left(t, \frac{n}{D}, R_A\right) + E_3\left(t, h, \frac{n}{D}\right) + E_4\left(t, u_2, h\right)$$

Where:

E_T = Evaporation from open surface of water (or equivalent in heat energy)

Δ =-Slope of vapor pressure curve at t = $\tan \alpha$

H = Equivalent evaporation of the total radiation on the surface of plants (the final amount - net- of energy finally remaining at a free water surface)

γ = psychrometer constant fixed device to measure humidity (= 0.66 if the temperature is measured ic centigrade and e in mbar)

E = Actual vapor pressure of air at temp t

E_a = Aerodynamic term (the term ventilation) depends on the air and low pressure steam (evaporation for the hypothetical case of equal temperatures of air and water)

t = $\tan \alpha$ = temperature

n/D = Cloudness ratio = (actual hours of sunshine) ÷ (possible hours of sunshine)

R_A = Agot's value of solar radiation arriving at the atmosphere (assuming no clouds and a perfectly transparent atmosphere).

h = Relative humidity

u_2 = Wind speed at height of 2 m (m/s)

Solution

1) Data: t = 22 °C, h = 0.6, sky Cover = 0.7, n ÷ D = 1 - sky cover Drag = 1 - 0.7 = 0.3, evaporation = 2.8 m/day, evapotranspiration = 0.8 E_o .
2) From nomogram for temperature t = 22° C & the value of n/D = 0.3 then, E_1 = - 2.2 mm/day
3) for Temperature t = 22 °C & tables for altitude 20° North In November, Agot's factor RA = 666 gm cal/cm^2/day & for values of n ÷ D = 0.3, E_2 = 2.6 mm/day

4) For temperature t = 22 °C, n ÷ D = 0.3, h = 0.6 then: E_3 = 1.2 mm/day

5) Potential evapotranspiration is given as = 2.8 mm/day
Since evapotranspiration = 0.8*evaporation, then evaporation predicted by Pennman is:
$0.8 \times E_o$ = 2.8 mm/day
Or E_o = 3.5 mm/day
E_o 3.5 = E_1 + E_2 + E_3 + E_4 = -2.2 + 2.6 + 1.2 + E_4
This gives E_4 = 1.9 mm/day

6) For a temperature t = 22 °C, E_4 = 1.9 mm/day humidity & h = 0.6, determine the wind speed at height of 2m, u_2 = 3.6 m/s then:

Example 4.2

Use the nomogram for the solution of Penman's equation to predict the daily potential evapotranspiration from a field crop at latitude 20°N in December, under the following conditions: (B.Sc., DU, 2012)

- Mean air temperature = 20°C
- Mean h = 70 %
- Sky cover = 60 %cloud
- Mean u_2 = 2.5 m/s
- Ratio of potential evapotranspiration to potential evaporation = 0.8

Solution

1. Data: latitude 20°N in December, t = 20° C, h = 0.7, sky Cover = 0.6, n ÷ D = 1 - sky cover Drag = 1 - 0.6 = 0.4, u_2 = 2.5 m/s, ET = 0.8E

2. From nomogram for temperature t = 20° C and the value of n/D = 0.4 then, E_1 = - 2.4 mm/day

3. for Temperature t = 20° C and tables for altitude 20° North In December, Agot's factor R_A = 599 gm cal/cm^2/day and for values of n ÷ D = 0.4 , E_2 = 2.6 mm/day

4. For temperature t = 20° C, h = 0.7, n ÷ D = 0.4, then: E_3 = 1.4 mm/day

5. For Temperature t = 20° C, speed u_2 = 2.5 m/s and humidity h = 0.7, then: E_4 = 1.1 mm/day

6. Find E_o from the equation:
   $E_o = E_1 + E_2 + E_3 + E_4 = -2.4 + 2.6 + 1.4 + 1.1 = 2.7$ mm/day
   Evapotranspiration = 0.7 E_o = 0.8 x 2.7 = 2.16 mm/day.

Program 4.2 Listing:

```
'*****************
'EXAMPLE 4.2
'*****************
Public Class Form1
    Dim Agot(10, 12) As Double
    Dim selAgotValue As Double

    Private Sub Form1_Load(ByVal sender As
        System.Object, ByVal e As
        System.EventArgs) Handles MyBase.Load

        Me.Text = "Example 4.2"
        Me.FormBorderStyle =
        Windows.Forms.FormBorderStyle.FixedDialog

        Label1.Text = "E1:"
        Label2.Text = "E2:"
        Label3.Text = "E3:"
        Label4.Text = "E4:"
        Label5.Text = "Ratio of potential
        evapotranspiration to potential
        evaporation:"
        Label6.Text = ""
        Button1.Text = "&Calculate"

        Label11.Text = "Latitude:"
        Label12.Text = "Month of year:"
        Label13.Text = ""
        Button2.Text = "&Find"
        GroupBox1.Text = "Find Agot's value:"

        'initialize Agot's values table
        init_Agot()
        init_Comboboxes()
    End Sub

    Private Sub init_Comboboxes()
      'this subroutine initializes the
      'comboboxes, which is a lengthy
```

88

```vb
'operation, albeit simple, that's why
'this code is placed in a separate sub,
'to make rest of the code clearer.
  ComboBox1.Items.Clear()
  ComboBox1.Items.Add("North 90")
  ComboBox1.Items.Add("North 80")
  ComboBox1.Items.Add("North 60")
  ComboBox1.Items.Add("North 40")
  ComboBox1.Items.Add("North 20")
  ComboBox1.Items.Add("Equator")
  ComboBox1.Items.Add("South 20")
  ComboBox1.Items.Add("South 40")
  ComboBox1.Items.Add("South 60")
  ComboBox1.Items.Add("South 80")
  ComboBox1.Items.Add("South 90")
  ComboBox1.SelectedIndex = 0

  ComboBox2.Items.Clear()
  ComboBox2.Items.Add("January")
  ComboBox2.Items.Add("February")
  ComboBox2.Items.Add("March")
  ComboBox2.Items.Add("April")
  ComboBox2.Items.Add("May")
  ComboBox2.Items.Add("June")
  ComboBox2.Items.Add("July")
  ComboBox2.Items.Add("August")
  ComboBox2.Items.Add("September")
  ComboBox2.Items.Add("October")
  ComboBox2.Items.Add("November")
  ComboBox2.Items.Add("December")
  ComboBox2.SelectedIndex = 0
End Sub

Private Sub init_Agot()
  'this subroutine initializes the Agot
  'array, which is a lengthy one, that's
  'why this code is placed in a
    'separate sub, to make the rest of
  'the code clearer.
  Agot(0, 0) = 0   'Values at N90
  Agot(0, 1) = 0
  Agot(0, 2) = 55
  Agot(0, 3) = 518
  Agot(0, 4) = 903
  Agot(0, 5) = 1077
  Agot(0, 6) = 944
  Agot(0, 7) = 605
  Agot(0, 8) = 136
  Agot(0, 9) = 0
  Agot(0, 10) = 0
```

```
Agot(0, 11) = 0
Agot(0, 12) = 3540
Agot(1, 0) = 0  'at N80
Agot(1, 1) = 3
Agot(1, 2) = 143
Agot(1, 3) = 518
Agot(1, 4) = 875
Agot(1, 5) = 1060
Agot(1, 6) = 930
Agot(1, 7) = 600
Agot(1, 8) = 219
Agot(1, 9) = 17
Agot(1, 10) = 0
Agot(1, 11) = 0
Agot(1, 12) = 3660
Agot(2, 0) = 86   'at N60
Agot(2, 1) = 234
Agot(2, 2) = 424
Agot(2, 3) = 687
Agot(2, 4) = 866
Agot(2, 5) = 983
Agot(2, 6) = 892
Agot(2, 7) = 714
Agot(2, 8) = 494
Agot(2, 9) = 258
Agot(2, 10) = 113
Agot(2, 11) = 55
Agot(2, 12) = 4850
Agot(3, 0) = 38     'at N40
Agot(3, 1) = 538
Agot(3, 2) = 663
Agot(3, 3) = 847
Agot(3, 4) = 930
Agot(3, 5) = 1001
Agot(3, 6) = 941
Agot(3, 7) = 843
Agot(3, 8) = 719
Agot(3, 9) = 528
Agot(3, 10) = 397
Agot(3, 11) = 318
Agot(3, 12) = 6750
Agot(4, 0) = 631    'at N20
Agot(4, 1) = 795
Agot(4, 2) = 821
Agot(4, 3) = 914
Agot(4, 4) = 912
Agot(4, 5) = 947
Agot(4, 6) = 912
Agot(4, 7) = 887
Agot(4, 8) = 856
```

```
Agot(4,  9) = 740
Agot(4, 10) = 666
Agot(4, 11) = 599
Agot(4, 12) = 8070
Agot(5,  0) = 844      'at Equator
Agot(5,  1) = 963
Agot(5,  2) = 878
Agot(5,  3) = 876
Agot(5,  4) = 803
Agot(5,  5) = 803
Agot(5,  6) = 792
Agot(5,  7) = 820
Agot(5,  8) = 891
Agot(5,  9) = 866
Agot(5, 10) = 873
Agot(5, 11) = 829
Agot(5, 12) = 8540
Agot(6,  0) = 970      'at S20
Agot(6,  1) = 1020
Agot(6,  2) = 832
Agot(6,  3) = 737
Agot(6,  4) = 608
Agot(6,  5) = 580
Agot(6,  6) = 588
Agot(6,  7) = 680
Agot(6,  8) = 820
Agot(6,  9) = 892
Agot(6, 10) = 986
Agot(6, 11) = 978
Agot(6, 12) = 8070
Agot(7,  0) = 998      'at S40
Agot(7,  1) = 963
Agot(7,  2) = 686
Agot(7,  3) = 515
Agot(7,  4) = 358
Agot(7,  5) = 308
Agot(7,  6) = 333
Agot(7,  7) = 453
Agot(7,  8) = 648
Agot(7,  9) = 817
Agot(7, 10) = 994
Agot(7, 11) = 1033
Agot(7, 12) = 670
Agot(8,  0) = 947      'at S60
Agot(8,  1) = 802
Agot(8,  2) = 459
Agot(8,  3) = 240
Agot(8,  4) = 95
Agot(8,  5) = 50
Agot(8,  6) = 77
```

```
    Agot(8, 7) = 187
    Agot(8, 8) = 403
    Agot(8, 9) = 648
    Agot(8, 10) = 920
    Agot(8, 11) = 1013
    Agot(8, 12) = 4850
    Agot(9, 0) = 981      'at S80
    Agot(9, 1) = 649
    Agot(9, 2) = 181
    Agot(9, 3) = 9
    Agot(9, 4) = 0
    Agot(9, 5) = 0
    Agot(9, 6) = 0
    Agot(9, 7) = 0
    Agot(9, 8) = 113
    Agot(9, 9) = 459
    Agot(9, 10) = 917
    Agot(9, 11) = 1094
    Agot(9, 12) = 3660
    Agot(0, 0) = 995      'at S90
    Agot(0, 1) = 656
    Agot(0, 2) = 92
    Agot(0, 3) = 0
    Agot(0, 4) = 0
    Agot(0, 5) = 0
    Agot(0, 6) = 0
    Agot(0, 7) = 0
    Agot(0, 8) = 30
    Agot(0, 9) = 447
    Agot(0, 10) = 932
    Agot(0, 11) = 1110
    Agot(0, 12) = 3540
End Sub

Private Sub Button1_Click(ByVal sender As
    System.Object, ByVal e As
    System.EventArgs) Handles Button1.Click

    Dim Eo, _E, E1, E2, E3, E4, ET As Double
    ET = Val(TextBox5.Text)
    E1 = Val(TextBox1.Text)
    E2 = Val(TextBox2.Text)
    E3 = Val(TextBox3.Text)
    E4 = Val(TextBox4.Text)
    Eo = E1 + E2 + E3 + E4
    _E = Eo * ET
    Label10.Text = "Eo = " + Eo.ToString + vbCrLf
    Label10.Text += "E = " + _E.ToString
End Sub
```

```
Private Sub Button2_Click(ByVal sender As
    System.Object, ByVal e As
    System.EventArgs) Handles Button2.Click

    'this sub will search for a value in Agot's
    'table according to the selected latitude
    'and month.
      If ComboBox1.SelectedIndex < 0 Then
         MsgBox("Please select a latitude from the
         option list.", vbOKOnly)
         Exit Sub
      End If
      If ComboBox2.SelectedIndex < 0 Then
         MsgBox("Please select a month from the
         option list.", vbOKOnly)
         Exit Sub
      End If

      selAgotValue = Agot(ComboBox1.SelectedIndex,
         ComboBox2.SelectedIndex)
      Label13.Text = "Agot="
      Label13.Text += selAgotValue.ToString
   End Sub
End Class
```

Example 4.3 (see Program 4.2 Listing)

Use the nomogram for the solution of Penman's equation to predict the daily potential evapotranspiration from a field crop at latitude 20°N in November, under the following conditions: (B.Sc., DU, 2012)

- Mean air temperature = $20^{\circ}C$
- Mean h = 80 %
- Sky cover = 70 %cloud
- Mean u_2 = 2.5 m/s
- Ratio of potential evapotranspiration to potential evaporation = 0.7

Solution

1) Data: sky Cover = 0.7, n ÷ D = 1 - sky cover Drag = 1 - 0.7 = 0.3, t = 20⬚C, h = 0.8, u_2 = 2.5 m/s, h = 0.8, Ratio of evapotranspiration to evaporation = 0.7.
2) From nomogram for temperature t = 20⬚C and the value of n/D = 0.3 then, E_1 = - 2.1 mm/day

93

3) For altitude 20⁰ North In November, from table find Agot's factor R_A = 666 gm cal/ cm²/day. Then, for Temperature t = 20⁰C and for values of n ÷ D = 0.3 and for R_A = 666 the value of E_2 = 2.3 mm/day

4) For temperature t = 20⁰C, h = 0.8 and n ÷ D = 0.3, then: E_3 = 1.2 mm/day

5) For Temperature t = 20⁰C, and the speed u_2 = 2.5 m / s and humidity h = 0.8, then: E_4 = 0.55 mm/day

6) Find E_o from the equation:
$E_o = E_1 + E_2 + E_3 + E_4$ = - 2.1 + 2.3 + 1.2 + 0.55 = 1.85 mm/day
Evapotranspiration = 0.7 Eo = 0.7 x 1.85 = 1.295 mm/day.

4.2 Theoretical Exercises

a) Outline importance of the evaporation process (B.Sc., DU, 2012).

b) Penman equation of evaporation from free water surfaces is based on two requirements that need to be met for continuous evaporation to occur: there must be a supply of energy to provide latent heat of vaporization, and there must be some mechanism for removing the vapor, once produced. Define the parameters shown in the equation (B.Sc., DU, 2012)

$$\frac{u}{u_o} = \left(\frac{z}{z_o}\right)^a$$

c) What the main differences between evaporation and transpiration?

d) What are Penman's assumptions to estimate amount of evaporation?

e) Explain the different factors affecting the evaporation Process.

4.3 Problem solving in evaporation

a) Use Penman nomogram to solve its equation to predict daily evapotranspiration expected from field plants at latitude 40 degrees north in April under the following conditions: intermediate air temperature 20 degrees Celsius, average humidity 70%, coverage of sky 60% clouds, relative speed at a height of two meters 2.5 m/s, and expected rate of transpiration to expected evaporation 0.7. What is the difference in the result at latitude of 40 degrees south? (Ans. 2.5 mm/day).

b) Use the Penman nomogram to solve its equation to predict expected daily evapotranspiration from field plants at latitude 60 degrees north in March and June, for an expected rate of evapotranspiration 65 percent of the expected evaporation, under the following conditions:

	March	June
Intermediate air temperature (° C)	5.5	15
Medium relative humidity (%)	79	77
Sky coverage (% clouds)	45	55
Relative velocity of wind (m/s)	2.5	3

c) Use the nomogram for the solution of Penman's equation to predict the daily potential evapotranspiration from a field crop at latitude 20°N in July, under the following conditions:
    Mean air temperature = 20°C
    Mean h = 80 %
    Sky cover = 70 %cloud
    Mean u2 = 2.5 m/s
    Ratio of potential evapotranspiration to potential evaporation = 0.8

d) Find amount of water evaporating from the surface of a lake during the month of March if the median value of the maximum temperature of the water surface of the lake during this period is 18 degrees Celsius, the temperature of dry air at a height of 2 meters above the lake surface is 20 degrees Celsius, humidity tw = 17 degree C, and the mean wind speed is 1.5 m/s at the same height noting that the lake did not freeze during the monitoring period.

e) Use the nomogram for the solution of Penman's equation to predict the daily potential evapotranspiration from a field crop at latitude 20°N in November, under the following conditions:
   • Mean air temperature = 20°C
   • Mean h = 80 %

- Sky cover = 70 %cloud
- Mean u_2 = 2.5 m/s
- Ratio of potential evapotranspiration to potential evaporation = 0.7

    (Ans. 1.295 mm/day).

Chapter Five

Infiltration and percolation

5.1 Infiltration

Infiltration is the movement of water through the soil surface into the soil. Percolation is the movement of water through the soil.

Factors affecting infiltration include: surface entry, transmission through soil, depletion of available storage capacity in soil and ccharacteristics of permeable medium.

Important indicators used to estimate infiltration (as an average rate throughout the period of storm and rain): Average infiltration method, φ-index, W-index and ratio to surface flow method.

Empirical models for infiltration capacity estimates relate infiltration rate or volume to elapsed time modified by certain soil properties. They include: Kostiakov model (irrigation applications), Horton equation (hydrologic modeling), Holton model (agriculture watersheds), Green-Ampt equation and Hydrologic Engineering Centre (HEC) of U.S. Army Corps of Engineers.

Φ - index is that rate of rainfall above which the rainfall volume equals the runoff volume. The index reflects the average infiltration rate, and is found from a time-rainfall intensity curve.

Example 5.1

The initial rate of infiltration of a watershed is estimated as 3 in/hr, the final capacity is 0.3 in/hr, and the time constant, k, is 0.4 per hour. Use Horton's Equation to find: (B.Sc., UoD 2014)

1. The infiltration capacity at t = 3 hr and t = 7 hr; and
2. The total volume of infiltration over the 7-hr period.

Solution

Horton's Equation: $f = f_c + (f_0 - f_c)e^{-kt}$

Where,

f = infiltration rate

f_0 = (initial) infiltration rate for dry ground = 3 in/hr

f_c = (asymptotic) infiltration rate for saturated ground = 0.3 in/hr

k = infiltration constant

Integrating Horton's equation over time gives the total depth of water that has infiltrated, F,

$$F = \int f dt = f_c t + \frac{(f_0 - f_c)}{k}[1 - e^{-kt}]$$

The infiltration capacity at t = 3 hr and t = 7 hr:

$f = f_c + (f_0 - f_c)e^{-kt}$

$f = 0.3 \text{ in/hr} + (3 - 0.3) \text{ in/hr} * e^{-0.4t} = 0.3 \text{ in/hr} + 22.7 \text{ in/hr} * e^{-0.4t}$

At t = 3 hr: $f = 0.3 \, in/hr + 2.7 \, in/hr * e^{-0.4*3} = 1.11 \, in/hr$

At t = 7 hr: $f = 0.3 \, in/hr + 2.7 \, in/hr * e^{-0.4*7} = 0.46 \, in/hr$

The total volume of infiltration over the 7-hr period

$$Volume = F = \int f dt = f_c t + \frac{(f_0 - f_c)}{k}[1 - e^{-kt}] = 0.3t + 2.7(e^{-0.4t})dt$$

The infiltration volume, F, is calculated as the integral of the infilrtration, f, equation, and it must be solved from time t = 0 to 7 hr. Simply plugging in t = 6 hr into the F equation is not complete.

$$Volume = \left[\left(0.3t + \frac{2.7}{0.4}\left(1 - e^{-0.4t}\right) \right) \right]_0^7$$

Program 5.1 Listing:

```
'***********************************
'EXAMPLE 5.1: Horton's equation
'***********************************
Public Class Form1

    Private Sub Form1_Load(ByVal sender As
        System.Object, ByVal e As System.EventArgs)
        Handles MyBase.Load

        Me.Text = "Example 5.1: Horton's equation"
        Me.FormBorderStyle =
            Windows.Forms.FormBorderStyle.FixedDialog

        Label1.Text =
             "Initial infiltration rate (in/hr)"
        Label2.Text =
             "Saturated ground infiltration rate"
        Label3.Text = "Infiltration constant, k"
        Label4.Text =
            "Find infiltration capacity after (hr)"
        Label5.Text = ""
        Button1.Text = "&Calculate"
    End Sub

    Private Sub Button1_Click(ByVal sender As
            System.Object, ByVal e As System.EventArgs)
            Handles Button1.Click
        Dim k, f, f0, fc, t As Double
        f0 = Val(TextBox1.Text)
        fc = Val(TextBox2.Text)
        k = Val(TextBox3.Text)
        t = Val(TextBox4.Text)

        f = fc + ((f0 - fc) * (Math.E ^ (-k * t)))
        Label5.Text = "At t=" + t.ToString + ", f="
                + FormatNumber(f, 2) + " in/hr"
    End Sub
End Class
```

99

Example 5.2

1) For a total rainfall distributed as shown in the table (1), the φ-index of the catchment area for a certain surface runoff is found to be 9 mm/hr (B.Sc., DU, 2012).

Table (1) Rainfall intensity versus time.

Time	Rainfall intensity (mm/hr)
0	0
1	5
2	15
3	20
4	20
5	14
6	1

Determine:

    i.   The value of surface runoff producing a φ–index of 9 mm/hr.

    ii.   The total rainfall in the catchment area.

2) If the same rainfall had been distributed as shown in table (2), compute the value of φ–index that would yield the same surface runoff. Comment on your answers.

Table (2) Rainfall intensity versus time.

Time	Rainfall intensity (mm/hr)
0	0
1	9
2	16
3	27
4	10
5	8
6	5

Solution

a) φ-index is the average rainfall intensity above which the volume of rainfall equals the volume of runoff.

b) Surface runoff = shaded area = (15-9)*1 + (20-9)*1 + (20-9)*1 + (14-9)*1 = 6*1 + 11* 1 + 11* 1 + 5*1 = 33 mm

Total rainfall = 5*1 + 15* 1 +20*1 + 20*1 + 14* 1 + 1* 1 = 75 mm

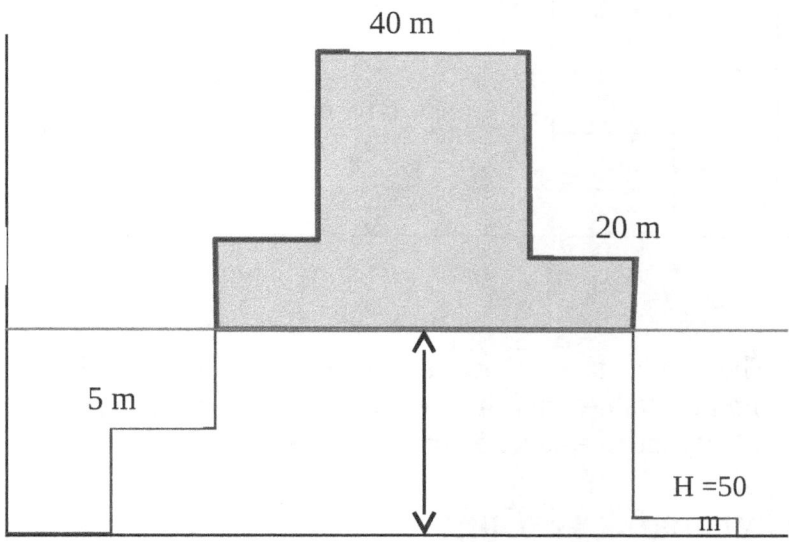

c)

Rate of rainfall above φ runoff volume

∴ (9 - φ)×1 + (16 - φ)×1 + (27 - φ)×1 + (10 - φ)×1 + (8 - φ)×1 = 33 mm

∴ 70 - 5φ = 33

∴ φ = 7.4 mm/hr

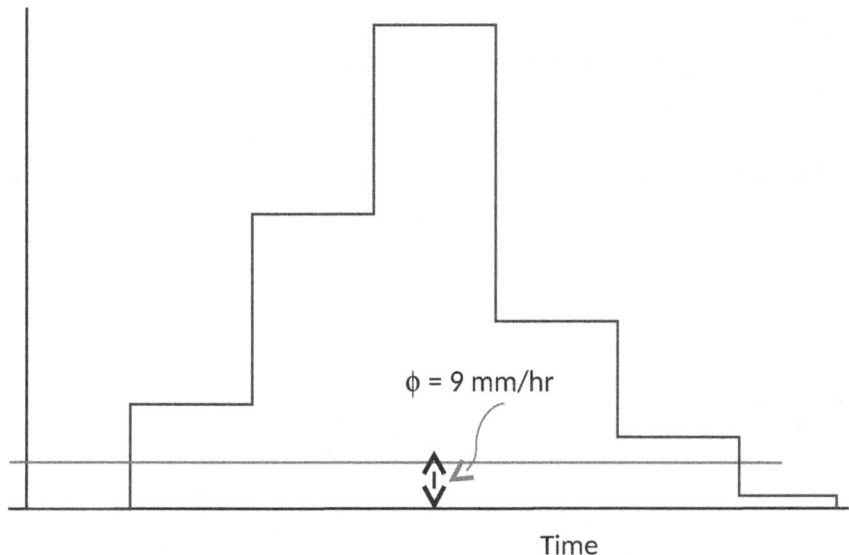

$\phi = 9$ mm/hr

Time

c) The results show that one determination of the f-index is of limited value and that many such determinations should be made, and averaged, before the index is used.

Program 5.2 Listing:

```
'************************
'EXAMPLE 5.2: Rainfall
'************************
Public Class Form1
    Private Sub Form1_Load(ByVal sender As
        System.Object, ByVal e As
        System.EventArgs) Handles MyBase.Load

        Me.Text = "Example 5.2: Rainfall"
        Me.FormBorderStyle =
        Windows.Forms.FormBorderStyle.FixedDialog
        Label1.Text = "Select option:"
        ComboBox1.Items.Clear()
        ComboBox1.Items.Add("Calculate surface runoff")
        ComboBox1.Items.Add("Calculate f-index")
        ComboBox1.SelectedIndex = 0
        Label3.Text = ""
        Button1.Text = "&Calculate"
        DataGridView1.Columns.Clear()
```

102

```
      DataGridView1.Columns.Add("rainCol",
   "Rainfall intensity (mm/hr)")
End Sub

Private Sub ComboBox1_SelectedIndexChanged(ByVal
   sender As System.Object, ByVal e As
   System.EventArgs) Handles
   ComboBox1.SelectedIndexChanged
   'change the user interface according
   'to the selected option
   Select Case ComboBox1.SelectedIndex
       Case 0
           Label2.Text = "Enter f-index (mm/hr):"
           TextBox1.Text = ""
       Case 1
           Label2.Text = "Enter surface
          runoff (mm):"
           TextBox1.Text = ""
   End Select
   DataGridView1.Rows.Clear()
End Sub

Private Sub Button1_Click(ByVal sender As
   System.Object, ByVal e As
   System.EventArgs) Handles Button1.Click

   Dim index, runoff, total, tmp As Double

   'sanity check -- if no data entered,
   'give an error message and exit gracefully.
   If DataGridView1.RowCount <= 1 Then
       MsgBox("Enter at least one value!",
      vbOKOnly)
       Exit Sub
   End If
   index = 0 : runoff = 0 : total = 0
   Select Case ComboBox1.SelectedIndex
       Case 0  'calculate surface runoff
           index = Val(TextBox1.Text)
           For i = 0 To DataGridView1.RowCount-2
             tmp =
      Val(DataGridView1.Rows(i).
         Cells(0).Value)
             If tmp > index Then
          runoff += (tmp - index)
             total += tmp
           Next
           Label3.Text = "Surface runoff = "
      + runoff.ToString + " mm"
           Label3.Text += vbCrLf
```

```vbnet
                        Label3.Text += "Total rainfall = "
            + total.ToString + " mm"
                '*****************************************
                '*****************************************
        Case 1   'calculate f-index
            Dim min, count As Double
            runoff = Val(TextBox1.Text)
            min =
Val(DataGridView1.Rows(0).
        Cells(0).Value)
            For i = 1 To DataGridView1.RowCount-2
                tmp =
Val(DataGridView1.Rows(i).
        Cells(0).Value)
                If min = 0 Then min = tmp
                If tmp < min Then min = tmp
            Next
            count = 0
            For i = 0 To DataGridView1.RowCount-2
                tmp =
Val(DataGridView1.Rows(i).
        Cells(0).Value)
                If tmp > min Then
                    total += tmp : count += 1
                End If
            Next
            index = (total - runoff) / count
            Label3.Text = "f-index = " +
        FormatNumber(index, 2) + " mm/hr"
    End Select
End Sub
End Class
```

Example 5.3 (see Program 5.2 Listing)

For a total rainfall of 44 mm, distributed as shown in the table (3), establish the φ index of the catchment area for a surface runoff of 19.5 mm. (B.Sc., DU, 2012)

Table (3) Rainfall intensity versus time.

Time	Rainfall intensity (mm/hr)
0	0
1	4
2	21
3	9
4	6
5	4

Solution

Rate of rainfall above φ runoff volume

∴ $(21 - φ)×1 + (9 - φ)×1 + (6 - φ)×1 = 19.5$ mm

∴ $36 - 3φ = 19.5$

∴ $φ = 5.5$ mm/hr

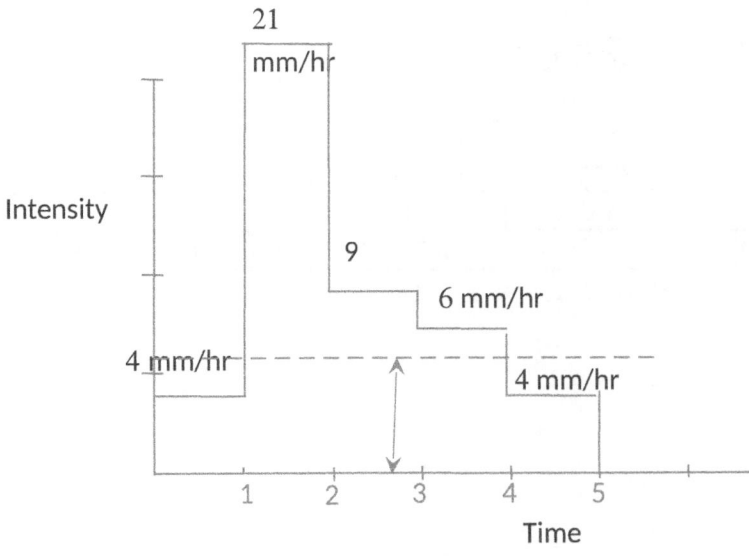

Example 5.4

A storm with 10.0 cm precipitation produced a direct runoff of 5.8 cm. Given the time distribution of the storm as below, estimate the φ-index of the storm.

Time from start (hr)	1	2	3	4	5	6	7	8
Incremental rainfall in each hour (cm)	0.4	0.9	1.5	2.3	1.8	1.6	1.0	0.5

Solution

Total Infiltration   = 10.0 cm - 5.8 cm = 4.2 cm
Assume time of rainfall excess, t_e = 8 hr   (for the first trial)
Then,

$$Index, \varphi = \frac{10.0 - 5.8}{8} = 0.525 \, cm/hr$$

But this value of φ makes the rainfalls of the first hour and eight hour ineffective as their magnitude is less than 0.525 cm/hr.
The value of t_e is therefore modified.
Assume time of rainfall excess, te = 6 hr   (for the second trial)
In this period, Infiltration = 10.0 - 0.4 - 0.5 - 5.8 = 3.3 cm

$$Index, \varphi = \frac{3.3}{6} = 0.55 \, cm/hr$$

This value of φ is satisfactory as it gives,
t_e = 6 hr, and by calculating rainfall excesses.

Time from start (h)	1	2	3	4	5	6	7	8
Rainfall Excess (cm)	0	0.35	0.95	1.75	1.25	1.05	0.45	0

Total Rainfall excess = 5.8 cm = total runoff

Program 5.4 Listing:

```vb
'*****************
'EXAMPLE 5.4
'*****************
Public Class Form1

    Private Sub Form1_Load(ByVal sender As
        System.Object, ByVal e As
        System.EventArgs) Handles MyBase.Load

        Me.Text = "Example 5.4"
        Me.FormBorderStyle =
        Windows.Forms.FormBorderStyle.FixedDialog

        Label1.Text = "Enter values for incremental
            rainfall each hour (cm):"
        Label2.Text = "Enter total runoff (cm):"
        Label3.Text = ""
        Button1.Text = "Estimate f-index"
        DataGridView1.Columns.Clear()
        DataGridView1.Columns.Add("rainCol",
        "Incremental hourly rainfall (cm)")
    End Sub

    Private Sub Button1_Click(ByVal sender As
        System.Object, ByVal e As
        System.EventArgs) Handles Button1.Click

        Dim runoff, index, N, totalRainFall As Double
        Dim totalInfiltration As Double

        If DataGridView1.RowCount <= 1 Then
            MsgBox("Enter at least one reading!")
            Exit Sub
        End If

        runoff = Val(TextBox1.Text)
        N = DataGridView1.RowCount - 1
        Dim tmp(N) As Double

        'compute total rainfall and load
        'rainfall readings into
        'an array to ease manipulation of
        'its values later.
        totalRainFall = 0
        For i = 0 To N - 1
            totalRainFall +=
    Val(DataGridView1.Rows(i).Cells(0).Value)
```

```
                tmp(i) =
    Val(DataGridView1.Rows(i).Cells(0).Value)
        Next
        totalInfiltration = totalRainFall - runoff
        index = totalInfiltration / N
check:
        If N = 0 Then
            Label3.Text =
    "No valid value of f-index found."
            Exit Sub
        End If

        'check the sanity of the calculation
        Dim tmp2 As Double = 0
        For i = 0 To DataGridView1.RowCount - 2
            If tmp(i) < index Then Continue For
            tmp2 += (tmp(i) - index)
        Next

        'if the calculation doesn't equal the
        'given runoff, remove the smallest of
        'values, recalculate the index
        'and infiltration, and recheck the sanity
        'check again.
        If tmp2 <> runoff Then
            For i = 0 To DataGridView1.RowCount - 2
                If tmp(i) < index And tmp(i) <> 0 Then
                    totalInfiltration -= tmp(i)
                    tmp(i) = 0
                    N -= 1
                    index = totalInfiltration / N
                    Exit For
                End If
            Next
            GoTo check
        End If
        Label3.Text = "t = " + N.ToString + vbCrLf
        Label3.Text += "f-index = " +
    FormatNumber(index, 2) + " cm/hr"
    End Sub
End Class
```

Example 5.5 (see Program 5.4 Listing)

Following data was obtained from a catchment of an area of 300 km^2.

Time from start (h)	1	2	3	4	5	6	7	8
Incremental rainfall in each hour (cm)	0.5	1.1	1.3	2.1	2.0	1.8	1.0	0.4

Solution

Estimate the ϕ index of the storm if the volume of surface runoff was 16.5 x 10^6 m^3.

Total Precipitation, P
   = 0.5+1.1+1.3+2.1+2.0+1.8+1.0+0.4
   = 10.2 cm

Runoff, R
   = (16.5 x 10^6 m^3) / (300 x 10^6 m^2)
   = 0.055 m = 5.5 cm

Total Infiltration, I   = P - R

Assume time of rainfall excess, te = 8 hr
   (for the first trial)

Then,

$$Index\,,\varphi = \frac{P-R}{t_e} = \ (10.2 - 5.5)\,/\,8 = 0.5875\ cm\,/\,hr$$

But this value of ϕ makes the rainfalls of the first hour and eight hour ineffective as their magnitude is less than 0.5875 cm/hr.

-   The value of te is therefore modified.
- Assume time of rainfall excess, te = 6 hr
   (for the second trial)

In this period, Infiltration =
In this period, Infiltration =

$$Index\,,\varphi = \frac{P-R}{t_e} =$$

This value of ϕ is satisfactory as it gives,
   te = 6 hr, and by calculating rainfall excesses.

Time from start (h)	1	2	3	4	5	6	7	8
Rainfall Excess (cm)	0	0.47	0.67	1.47	1.37	1.17	0.37	0

Total Rainfall excess = 5.5 cm = total runoff

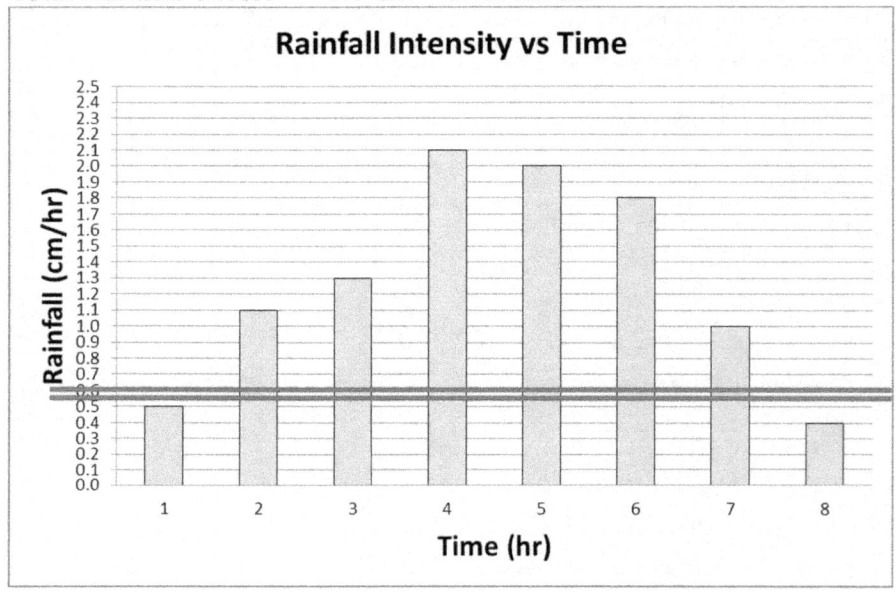

5.2 Antecedent precipitation index, API
Example 5.6

The API for a station was 39 mm on first December, 61 mm rain fell on fifth December, 42 mm on 7 December & 19 mm on 8 December. Compute the anticedent precipitation index for 12 December, if k = 0.89, and for same data assuming no rain fell (B.Sc., DU, 2013).

$$I_i = I_o * k^t$$

Where:

I_t = Index value at t days later (mm)

I_o = Initial value to the index (mm)

k = A recession constant, of magnitude between 0.85 to 0.98 (usually taken for a value of about 0.92).

t = Time (day).

Solution

- Data: amount of precipitation in different days of the month of December.
- Since $I_t = I_o k^t$ & assuming t = 1 produces: It = $I_o k$. Which means that: the index for any day is equal to that value in the previous day multiplied by the coefficient k. then if rain fell in the day, the amount of rain is added to the index.
- For the given data index can be estimated as follows: On the first day of December, $I_1 = 39$ mm Just on the fifth day of December $I_5 = 39*(0.89)^4$, $I_5 = 24.5$ mm
- On the fifth day of December rain should be added: $I_5 = 24.5 + 61 = 85.5$mm
- Just on the seventh day of December $I_7 = 85.5* (0.89)^2$, $I_7 = 67.7$ mm
- On the seventh day of December rain should be added : $I_7 = 67.7 + 42 = 109.7$mm
- Just on the eighth day of December $I_8 = 109.7*(0.89)^1$, $I_8 = 97.6$ mm
- On the eighth day of December rain should be added: $I_8 = 97.6 + 19 = 116.6$ mm
- In the twelfth day of December a $I_{12} = 116.6*(0.89)^4$, $I_{12} = 73.2$ mm (with rain in different days)
- In the case of lack of rain, the value is equal to the directory: $I_{12} = 39*(0.89)^{11}$, $I_{12} = 10.8$ mm (with no rain)

Program 5.6 Listing:

```
'*****************
'EXAMPLE 5.6: API
'*****************
Public Class Form1

    Private Sub Form1_Load(ByVal sender As
        System.Object, ByVal e As
        System.EventArgs) Handles MyBase.Load

        Me.Text = "Example 5.6: API"
        Me.FormBorderStyle =
        Windows.Forms.FormBorderStyle.FixedDialog
```

```
        Label1.Text =
    "Enter daily rainfall values (mm):"
    Label2.Text = "Recession constant, k:"
    Label3.Text = "Estimate rainfall on day:"
    Label4.Text = ""
    Button1.Text = "Estimate rainfall"
    DataGridView1.Columns.Clear()
    DataGridView1.Columns.Add("dayCol",
    "Day of month")
    DataGridView1.Columns.Add("rainCol",
    "Daily rainfall (mm)")
End Sub

Private Sub Button1_Click(ByVal sender As
    System.Object, ByVal e As
    System.EventArgs) Handles Button1.Click

    Dim k, N As Double
    Dim I(N), t, t2, tmp, tmp2 As Double

    If DataGridView1.RowCount <= 1 Then
        MsgBox("Enter at least one reading!")
        Exit Sub
    End If

    k = Val(TextBox1.Text)
  'number of rainy days
    N = DataGridView1.RowCount - 1
  I(0) =
    Val(DataGridView1.Rows(0).Cells(1).Value)
    t = Val(DataGridView1.Rows(0).Cells(0).Value)
    'formula: Ix = Iy*k^t.. here t is
    'the difference between
    'day 'x' and day 'y' readings were taken.
    For j = 1 To N - 1
        tmp = I(j - 1) * (k ^
        (Val(DataGridView1.Rows(j).
        Cells(0).Value) - t))
        I(j) = tmp +
        Val(DataGridView1.Rows(j).
        Cells(1).Value)
        t = Val(DataGridView1.Rows(j).
        Cells(0).Value)
    Next
    t = Val(DataGridView1.Rows(0).Cells(0).Value)
    'see what day is requested
    t2 = Val(TextBox2.Text)
    'calculate I depending on rain calculated
    'in the last day
    tmp = I(N - 1) * (k ^ (t2 -
```

112

```
        Val(DataGridView1.Rows(N - 1).
            Cells(0).Value)))
        'calculate I depending on rain calculated
        'for first day
        tmp2 = I(0) * (k ^ (t2 - t))
        'output the results
        Label4.Text = "I" + t2.ToString + "=" +
        FormatNumber(tmp, 2) + _
            " mm (with rainy days)" + vbCrLf
        Label4.Text += "I" + t2.ToString + "=" +
        FormatNumber(tmp2, 2) + _
            " mm (without rain)"
    End Sub
End Class
```

5.3 Theoretical Exercises

1) The antecedent precipitation index (API) is given by: $I_e = I_o k^t$, Define the parameters shown in the equation (B.Sc., UAE, 1989, B.Sc., DU, 2012).

2) Define ϕ-index. (B.Sc., DU, 2012)

3) Illustrate major factors that may affect infiltration of water within a certain catchment area. (B.Sc., UAE, 1989).

4) What are the factors affecting infiltration of water into the soil. How is it measured?

5) What is the difference between infiltration and percolation? Indicate how both are measured.

6) Elaborate on indicators used to estimate infiltration. Which one is the best to be used? Why?

7) Define infiltration. What are the factors that govern infiltration of water through the soil? (B.Sc., UoD 2014)

5.4 Problem solving in infiltration

ϕ-index

1. Find value of ϕ-index in a catchment area to make a surface runoff of 19.5 mm, knowing that the total value of the rain for an area equal to 44 mm as indicated in the following table: (Ans. 5.5 mm/hr).

113

Time, hr	Rainfall intensity, mm/hr
0	0
1	4
2	21
3	9
4	6
5	4

2. For a total rainfall of 150mm, distributed as shown in table (2), establish the φ index of the catchment area for a surface runoff of 72mm. (B.Sc., UAE, 1989).

Time	Rainfall intensity (mm/hr)
0	0
1	16
2	29
3	35
4	16
5	8
6	19
7	27

Compute the change in φ index for the same catchment in (b) above if a second storm gave rise to a runoff equivalent to 68mm while the hourly rainfall of the storm indicated the following distribution: (B.Sc., UAE, 1989). (ans. 11.7 mm/hr,).

Time	Rainfall intensity (mm/hr)
0	0
1	15
2	43
3	61
4	24
5	12

3. Determine the φ index for the catchment area as affected by the two storms indicated in (a) and (b) above. (B.Sc., UAE, 1989). (Ans. 11.7, 20 mm/hr)
4. Use the rainfall data below to determine the φ index for the watershed if the runoff depth was 6.6 in.

Time, hr	Rainfall, in/hr
0 - 1	2.3
1 - 3	3.1
3 - 5	3.3
5 - 8	2.6

API

5. The API for a station was 60 mm on the first of December, 64 mm rain fell on the seventh of December, 40 mm rain fell on the ninth of December, and 30 mm rain fell on the eleventh of December. Compute the API for 20th December if k = 0.85. Determine the API if no rain fell (B.Sc., DU, 2012). (Ans. 24.1, 2.7 mm)

6. The API for a station was 55mm on the first of August, 57mm rain fell On the sixth of August, 31mm rain fell on the eighth of August, and 20mm rain fell on the 9th August. Compute the API for 15th August if k = 0.92. Determine the APU if no rain fell. (B.Sc., UAE, 1989). (Ans. 73.5, 17.1 mm).

7. The value of API for rainfall monitoring stations reached 42 mm in the first day of October, and an amount of 46 mm of rain fell in the fifth day of October, and an another amount of rain of 28 mm fell on the seventh day and 34 mm fell on the eighth day of October. Find the value of the API on 12 October assuming that k is equal to 0.92. Calculate the amount of index assuming no rain fell in the same period.

8. The API for a station was 55mm on the first of August, 57mm rain fell On the sixth of August, 31mm rain fell on the eighth of August, and 20mm rain fell on the 9th August. Compute the API for 15th August if k = 0.92. Determine the API if no rain fell. (ans. 17.1 mm).

9. The value of the API for a certain station reached 42 mm in first of August, and an amount of 46 mm of rain fell on the fifth of August, also rain of 28 mm fell in the seventh of august and 34 mm on the eighth day of August. Find value of the API guide on 12th August noting that k is equal to 0.92. Calculate the amount of the index assuming no rain in the

same period. Draw curve of change of index with time. (Ans. 85.3, 1.7 mm).

10. The value of the API for a station reached 55 mm in the first of August, an amount of 57 mm of rain fell on the sixth of August, 31 mm of rain fell in the eighth of August and 20 mm fell in the ninth day of August. Find value of API index for the 15th day of August, noting that k is equal to 0.92. Calculate the amount of index assuming no rain during the same period. Draw curve of change of index with time.

11. The antecedent precipitation index (API) is given by: Ie = Iokt, Define the parameters shown in the equation. The API for a station was 60 mm on the first of December, 64 mm rain fell on the seventh of December, 40 mm rain fell on the ninth of December, and 30 mm rain fell on the eleventh of December. Compute the API for 20th December if k = 0.85. Determine the API if no rain fell. (ans. 24.1, 2.7 mm).

Chapter Six

Groundwater Flow

6.1 Groundwater

Groundwater represents that part of reserved water in a porous aquifer, resulting from infiltration of rainfall through the soil and its penetration to the underlying strata.

Example 6.1

Outline most important factors affecting productivity of a well (B.Sc., DU, 2013).

Solution

Most important factors affecting productivity of a well:

- Lowering of groundwater within aquifer (drawdown aspects).
- Dimensions of aquifer & its lateral extent.
- Ground water storage.
- Transmissivity & specific yield or storage coefficient of aquifer.
- Conditions of flow (steady or unsteady).
- Depth of well.
- Establishment of well & methods of construction, properties & condition.

Table (6.1) Expected well yield.

Well diameter, cm	Expected well yield, m^3/d
15	< 500
20	400 to 1000
25	800 to 2000
30	2000 to 3500
35	3000 to 5000
40	4500 to 7000
50	6500 to 10000
60	8500 to 17000

Example 6.2

A well is 30 cm diameter and penetrates 50 m below the static water table. After 36 hr of pumping at 4.0 m^3/minute the water level on a test well 200 m distance is lowered by 1.2m and in a well 40 m away the drawdown is 2.7 m (B.Sc., DU, 2013).

1. Determine radius of zero draw-down.
2. Find coefficient of permeability.
3. Compute draw-down in the pumped well.
4. What is the transmissibility of the aquifer?

Using expected well yield estimates as presented in table (5.1), comment about yield of the well as related to its diameter, & suggest a more suitable well diameter. Explain and validate your answer (B.Sc., DU, 2013).

$$Q=\frac{\pi k\left(H^2-h^2\right)}{Ln\frac{R}{r}}$$

Solution

a) Given: D = 0.3 m, H = 50 m, r_1 = 40 m, x_1 = 2.7 m, r_2= 200 m, x_2= 1.2 m, Q_o = 4000 L/min.

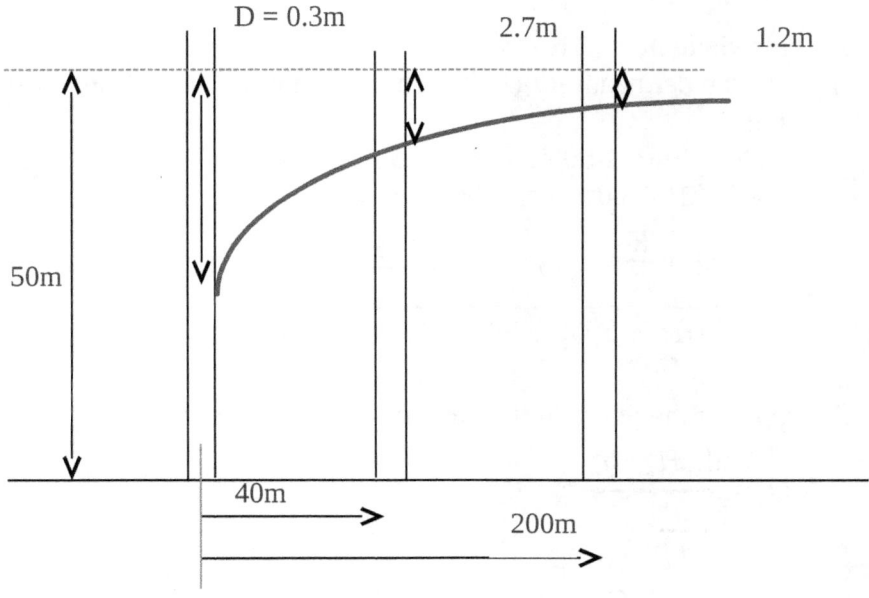

Find h_1 = h - x_1 = 50 – 2.7 = 47.3 m, and h_2 = h - x_2 = 50 – 1.2 = 48.8 m.

Use the following equation for both observation wells:

$$Q=\frac{\pi k\left(H^2-h^2\right)}{Ln\dfrac{R}{r}}$$

Where:

Q = Flow through well at a distance r.

h = Depth below original water level.

R = radius of zero draw down.

k = coefficient of permeability.

H = depth of aquifer.

$$\therefore \left[\frac{\pi k\left(H^2-h^2\right)}{\ln\dfrac{R}{r}}\right]_{1st\ well} = \left[\frac{\pi k\left(H^2-h^2\right)}{\ln\dfrac{R}{r}}\right]_{2nd\ well}$$

By substituting given values into the previous equation, then:

$$\therefore \left[\frac{\left(50^2-47.3^2\right)}{\ln\dfrac{R}{40}}\right]_{\text{ist well}} = \left[\frac{\left(50^2-48.8^2\right)}{\ln\dfrac{R}{200}}\right]_{\text{2nd well}}$$

b) This yields R = 751.5 m.

c) Find the permeability coefficient by using the data of one of the wells.

Thus, for h = 50 m, ho = 47.3 m, r = 40 m, R = 751.5 m, Q = 4000*10⁻³*60*24 = 5760 m³/day,

$$k=\frac{QLn\dfrac{R}{r_1}}{\pi\left(H^2-h_1^2\right)}=\frac{5760Ln\dfrac{751.5}{40}}{\pi\left(50^2-47.3_1^2\right)}=20.5 \ \text{m/d}$$

d) Depth of the water in the pumped well may be found as:

$$Q_o=\frac{\pi k\left(H^2-h_o^2\right)}{Ln\dfrac{R}{r_o}} \qquad \text{Or}$$

$$h_0^2=H^2 \ - \ \frac{Q}{\pi k}Ln\frac{R}{r_o}=50^2 \ - \ \frac{5760}{\pi x20.5}Ln\frac{751.5}{0.15}$$

This yields, h = 41.7 m

Determine the draw-down at the well as:

r₀ = h - h₁ = 50 – 41.7 = 8.3 m.

e) From table for a well diameter of 30 cm, yield is 2000 – 3500 m³/d.

The amount of water abstracted is Q₀ = 4000 L/min = 4000*60*24/1000 = 5760 m³/d. A better design would be selected a well of diameter of 40 cm (giving a yield of 5867 m³/d for computed drawdown sat well (O.k. between 4500 - 7000).

$$Q_o=\frac{\pi\times20.5\left(50^2-41.7^2\right)}{Ln\dfrac{751.5}{(0.4/2)}}=5867 \ \text{cubic meter per day}$$

Program 6.1 Listing:

```
'*****************************
'EXAMPLE 6.1: Groundwater I
'*****************************
Public Class Form1

    Private Sub Form1_Load(ByVal sender As
        System.Object, ByVal e As
        System.EventArgs) Handles MyBase.Load

        Me.Text = "Example 6.1"
        Me.FormBorderStyle =
        Windows.Forms.FormBorderStyle.FixedSingle
        Me.MaximizeBox = False
        Label1.Text = "Well's diameter, d (m)"
        Label2.Text = "Well's depth, H (m)"
        Label3.Text = "r1 (m)"
        Label4.Text = "x1 (m)"
        Label5.Text = "r2 (m)"
        Label6.Text = "x2 (m)"
        Label7.Text = "Qo (L/m)"
        Label8.Text = ""
        Button1.Text = "&Calculate"
    End Sub

    Private Sub Button1_Click(ByVal sender As
        System.Object, ByVal e As
        System.EventArgs) Handles Button1.Click

        Dim diameter, H, r1, x1, r2, x2, Qo, h1,
            h2 As Double
        Dim R, a, b, c, d, k, h0 As Double

        diameter = Val(TextBox1.Text)
        H = Val(TextBox2.Text)
        r1 = Val(TextBox3.Text)
        x1 = Val(TextBox4.Text)
        r2 = Val(TextBox5.Text)
        x2 = Val(TextBox6.Text)
        Qo = Val(TextBox7.Text)
        h1 = H - x1
        h2 = H - x2

        'calculate R
        a = (H ^ 2) - (h1 ^ 2)
        b = Math.Log(r1)
        c = (H ^ 2) - (h2 ^ 2)
        d = Math.Log(r2)
```

121

```
        R = Math.E ^ (((a * d) - (c * b)) / (a - c))
        Label8.Text = "R = " + FormatNumber(R, 2) +
            " m"
        'find permeability coefficient
        Qo = Qo / 1000 * 60 * 24
        k = (Qo * Math.Log(R / r1)) /
        (Math.PI * ((H ^ 2) - (h1 ^ 2)))
        Label8.Text += vbCrLf
        Label8.Text += "k = " + FormatNumber(k, 2) +
            " m/d"
        Label8.Text += vbCrLf
        'find depth of pumped well
        h0 = Math.Sqrt((H ^ 2) -
        ((Qo / (Math.PI * k)) *
        Math.Log(R / (diameter / 2))))
        Label8.Text += "Water depth, h1 = " +
        FormatNumber(h0, 2) + " m"
        Label8.Text += vbCrLf
        Label8.Text += "Drawdown at well = " +
        FormatNumber(H - h0, 2) + " m"
    End Sub
End Class
```

Example 6.2

A fully-penetrating well, with an outside diameter of 50 cm, discharges a constant 3 m^3/min from an aquifer whose coefficient of transmissibility is 0.03 m^2/s. The aquifer is in contact with a lake 2 km away & has no other source of supply (B.Sc., DU, 2012).

1. Estimate the drawdown at the well surface. (Take R_0 as twice distance between aquifer & lake).
2. Using expected well yield estimates as presented in table (1), comment about yield of the well as related to its diameter, & suggest a more suitable well diameter. Explain your answer.
3. Give a suggestion for type of aquifer soil with assuming an aquifer thickness of 20 m.

Solution

1) Data: Q_0 = 3 m^3/min = 3*60*24 = 5760 m^3/day, R_0 = 2xL = 2x2000 = 4000m, D = 0.5 m, r_0 = 0.5/2 = 0.25 m, T = 0.03 m^3/s = 1.8 m^2/min
2) Using equation

3)

$$S_o = \frac{Q_o}{2\pi kH} \ln \frac{R_o}{r} = \frac{Q_o}{2\pi T} \ln \frac{R_o}{r} \qquad S_o = \frac{3}{2\pi \times 1.8} \ln \frac{4000}{0.25}$$

Hence, drawdown at well face is 2.57 m

- From table of well yield this diameter (= 0.5m) gives a yield in range of 6500 – 10000 m^3/d. Nonetheless, needed constant flow ought to be 4320 m^3/d. This suggests using a well of diameter of 35 cm to have a yield between 3000 to 5000 m^3/d, thus satisfying need.
- The actual well yield according to aquifer properties and computed drawdown for a well diameter of 35 cm would then be:

$$Q_o = \frac{2\pi T}{Ln \frac{R}{r_o}} S_o = \frac{2\pi \times 1.8 * 60 * 24 * 2.57}{Ln \frac{3000}{0.175}}$$

$$= 4167 \ m^3/d \ \ between \ 3000 \ and \ 5000 \ O.K.$$

- Determine permeability k = T/H = (1.8*60*24)/20 = 129.6 m/d
- From figure of range of permeability in natural soils, this value of k suggests **fine gravel** soil description.

Program 6.2 Listing:

```
'****************************
'EXAMPLE 6.2: Groundwater II
'****************************
Public Class Form1
    Dim well_yield_diam(7) As Integer
    Dim well_yield_min(7) As Integer
    Dim well_yield_max(7) As Integer
    Const TABLE_TOTAL = 8

    Private Sub Form1_Load(ByVal sender As
        System.Object, ByVal e As
        System.EventArgs) Handles MyBase.Load

        Me.Text = "Example 6.2"
        Me.FormBorderStyle =
        Windows.Forms.FormBorderStyle.FixedSingle
```

```vb
        Me.MaximizeBox = False
        Label1.Text = "Diameter, D (m)"
        Label2.Text = "Discharge rate, Q (m3/min)"
        Label3.Text = "Coefficient, T (m2/s)"
        Label4.Text = "R0 (m)"
        Label5.Text = ""
        Button1.Text = "&Calculate"

        'initialize table of expected well yields
        well_yield_diam(0) = 15
        well_yield_diam(1) = 20
        well_yield_diam(2) = 25
        well_yield_diam(3) = 30
        well_yield_diam(4) = 35
        well_yield_diam(5) = 40
        well_yield_diam(6) = 50
        well_yield_diam(7) = 60
        well_yield_min(0) = 0
        well_yield_min(1) = 400
        well_yield_min(2) = 800
        well_yield_min(3) = 2000
        well_yield_min(4) = 3000
        well_yield_min(5) = 4500
        well_yield_min(6) = 6500
        well_yield_min(7) = 8500
        well_yield_max(0) = 500
        well_yield_max(1) = 1000
        well_yield_max(2) = 2000
        well_yield_max(3) = 3500
        well_yield_max(4) = 5000
        well_yield_max(5) = 7000
        well_yield_max(6) = 10000
        well_yield_max(7) = 17000
End Sub

Private Sub Button1_Click(ByVal sender As
    System.Object, ByVal e As
    System.EventArgs) Handles Button1.Click

        Dim D, Q, T, R0, r, S As Double
        D = Val(TextBox1.Text)
        Q = Val(TextBox2.Text)
        T = Val(TextBox3.Text)
        R0 = Val(TextBox4.Text)
        'Q *= (60 * 24)
        T *= 60
        r = D / 2
        S = (Q / (2 * Math.PI * T)) *
        (Math.Log(R0 / r))
        Label5.Text = "Draw at well face = " +
```

```
        FormatNumber(S, 2) + " m"

    D *= 100    'convert to cm

    'find expected well yield in the
    'table using diameter as index
    For i = 0 To TABLE_TOTAL - 1
        If well_yield_diam(i) = D Then
            Label5.Text += vbCrLf
            Label5.Text +=
        "From table, diameter gives a
        yield in range " _
        + vbCrLf + "of" +
        well_yield_min(i).ToString + " - "
                + well_yield_max(i).ToString
            Exit For
        End If
    Next
  End Sub
End Class
```

Example 6.3 (see Program 6.2 Listing)

A fully-penetrating well, with an outside diameter of 0.3 m, discharges a constant 4 m^3/min from an aquifer whose coefficient of transmissibility is 1.4 m^2/min. The aquifer is in contact with a lake 1.5 km away & has no other source of supply(B.Sc., DU, 2012).

1. Estimate the drawdown at the well surface. (Take R_0 as twice distance between aquifer & lake).
2. Using expected well yield estimates as presented in table (1), comment about yield of the well as related to its diameter, & suggest a more suitable well diameter. Explain your answer.
3. Give a suggestion for type of aquifer soil with assuming an aquifer thickness of 25 m.

Solution

1. Theim's equation assumptions
    - Aquifer homogeneous, isotropic & extended to infinity.
    - Well penetrates thickness of aquifer carrying water & diverting water from it.
    - Transmissivility (hydraulic conductivity) is constant in each place & does not depend on time.

125

- Pumping is conducted at a steady rate for a period so that it may assume a stable condition.
- Stream lines are radial (horizontal).
- Flow is laminar.

2. Data: Q_0 = 4 m³/min = 4*60*24 = 5760 m³/day, R_o = 2xL = 2x1500 = 3000m, D = 0.3 m, r_o = 0.3/2 = 0.15 m, T = 1.4 m²/min

3. Using equation

4.

$$S_o = \frac{Q_o}{2\pi kH}\ln\frac{R_o}{r} = \frac{Q_o}{2\pi T}\ln\frac{R_o}{r} \qquad S_o = \frac{4}{2\pi\times 1.4}\ln\frac{3000}{0.15}$$

Hence, drawdown at well face is 4.5 m

- From table of well yield this dia (= 0.3m) gives a yield in range of 2000 – 3500 m³/d. Nonetheless, needed constant flow ought to be 5760 m³/d. This suggests using a well of diameter of 40 cm to have a yield between 4500 to 7000 m³/d, thus satisfying need.
- The actual well yield according to aquifer properties and computed drawdown for a well diameter of 40 cm would then be:

$$Q_o = \frac{2\pi T}{Ln\frac{R}{r_o}}S_o = \frac{2\pi\times 1.4*60*24*4.5}{Ln\frac{3000}{0.2}} = 5929\ m^3/d > 5760\ O.K.$$

- Determine permeability k = T/H = (1.4*60*24)/25 = 80.6 m/d
- From figure of range of permeability in natural soils, this value of k suggests **coarse sand** soil description.

Example 6.4 (see Program 6.1 Listing)

The static level of water table in an unconfined aquifer was 30 m above the underlying impermeable stratum. A 150 mm diameter well, penetrating the aquifer to its full depth, was pumped at the rate of 20 litres per second. After several weeks of pumping, the drawdown in observation wells 20 m and 50 m from the well were 3.5 m and 2 m respectively, and the observed drawdowns were increasing very slowly (B.Sc., DU, 2012).

a) Assuming equilibrium conditions, estimate the hydraulic conductivity and transmissivity of the aquifer.
b) Estimate the drawdown just outside the pumped well.
c) What will be the yield of a 300 mm diameter well which will produce the same drawdowns just outside the well and at the 50 m distance observation well in (2)? What would be the drawdown at a nearer observation well?

Solution

Determine Q = 20x10^{-3}x60x60x24 = 1728 m^3/d

$$Q=\frac{\pi k \left(H^2-h^2\right)}{\ln \frac{R}{r}}$$

$$\left[\frac{\pi k\left(H^2-h^2\right)}{\ln \frac{R}{r}}\right]_{1\,st\,well}=\left[\frac{\pi k\left(H^2-h^2\right)}{\ln \frac{R}{r}}\right]_{2\,nd\,well}$$

$$\frac{30^2-\left(30-3.5\right)^2}{\ln \frac{R}{20}}=\frac{30^2-\left(30-2\right)^2}{\ln \frac{R}{50}}$$

∴ R = 183.5 m

$$k=\frac{Q.\ln \frac{R}{r}}{\pi\left(H^2-h^2\right)}=\frac{1728.\ln \frac{183.5}{20}}{\pi\left[30^2-\left(30-3.5\right)^2\right]}$$

$$= 6.16 \text{ m/d} = 7.13 \times 10^{-5} \text{ m/s}$$

T = kH = 6.16×30 = 184.9 m^2/d

b) Depth of water in well,

$$h^2=H^2-\frac{Q}{\pi k}\ln \frac{R}{r}=30^2-\frac{1728}{\pi \times 6.16}\ln \frac{183.5}{0.075} \quad = 203.8$$

H = 14.3 m

Drawdown just outside the = 30 − 14.3 = 15.7 m.

c) Yield

$$Q=\frac{\pi k\left(H^2-h^2\right)}{\ln\frac{R}{r}}$$

$$=\frac{\pi \times 6.16\left[30^2-(30-2)^2\right]}{\ln\frac{183.5}{0.15}}=315.8\frac{m^3}{d}=3.66 \times 10^{-3}\frac{m^3}{s}$$

$$h^2=30^2-\frac{315.8}{\pi \times 6.16}\ln\frac{183.5}{0.15}=784$$

h = 28 m.

Drawdown at a nearer observation well = 30 − 28 = 2 m

Example 6.5 (see Program 6.1 Listing)

A well of diameter 30 cm penetrates an aquifer, water depth in it 15 meters before pumping. When pumping is being done at a rate of 3000 liters per minute, the drawdowns in two observation wells 40 and 90 meters away from the well are found to be 1.3 and 0.7 meters, respectively (B.Sc., DU, 2011).

1. Find radius of zero draw down.
2. Determine the coefficient of permeability, and
3. Compute drawdown in the well.
4. Give a suggestion for type of aquifer soil for the estimated permeability.
5. Comment on yield of well compared to its diameter. Give suggestions for improvement?

Solution

1) Data: Data on wells.
2) Use following equation for each well
   r = 40 m, s = 1.3 m, h = 15 − 1.3 = 13.7 m
   r = 90 m, s= 0.7m, h = 15 − 0.7 = 14.3 m

$$Q_o=\frac{\pi k\left(H^2-h_0^2\right)}{Ln\frac{R}{r_o}}$$

128

$$\therefore \left[\frac{\pi k \left(H^2 - h_0^2\right)}{Ln\dfrac{R}{r_o}}\right]_{ist\,well} = \left[\frac{\pi k \left(H^2 - h_0^2\right)}{Ln\dfrac{R}{r_o}}\right]_{2nd\,well}$$

$$\left[\frac{\left(15^2 - 13.7^2\right)}{Ln\dfrac{R}{40}}\right]_1 = \left[\frac{\left(15^2 - 14.3^2\right)}{Ln\dfrac{R}{90}}\right]_2$$

i.   Then: R can be found = 242.2 m

ii.   Find coefficient permeability from the equation to the data of the well about 40 m away: H = 15 m, s = 1.3 m, h = 15 - 1.3 = 13.7 m, R = 254.9 m, Q = 3000 L/min = 3 m³/min = 3*60*24 = 4320 m³/day

$$k = \frac{QLn\dfrac{R}{r_o}}{\pi\left(H^2 - h_0^2\right)} = \frac{4320\,Ln\dfrac{242.29}{40}}{\pi\left(15^2 - 13.7^2\right)} = 66.36\,m/d$$

iii.   To find the depth of the water in the well, use equation with $r_0 = 0.3/2 = 0.15$ and k = 66.36 m/d:

$$h_0^2 = H^2 - \frac{Q}{\pi k}Ln\frac{R}{r} = 15^2 - \frac{1576.8}{\pi x\,24.9}Ln\frac{254.9}{0.15} = 75.09$$

      From which: h = 8.7 m

      Then drop in level = drawdown at well face = s_0 = 15 − 8.7 = 6.3 m

iv.   Suggest type of aquifer natural soil for a permeability of 66.36 m/day from given diagram to be "coarse sand".

v.   yield of well for a diameter of 30 cm ought to be between 2000 to 3500 m3/d. Pumping done at a rate of 4320 m3/d which suggests a suitable diameter of 35 cm. Taking this diameter would yield a flow of:

$$Q_o = \frac{\pi k \left(H^2 - h_0^2\right)}{Ln\dfrac{R}{r_o}} = \frac{\pi x\,66.36\left(15^2 - 8.48^2\right)}{Ln\dfrac{242.2}{0.15}} = 4322\,m\,3/d$$

Thus a well diameter of 35 cm is suggested for improvement to yield a flow rate of 4322 m3/d (within yield of 3000 to 5000, O.K.).

Example 6.6

A catchment area is undergoing a prolonged rainless period. The discharge of the stream draining it is 140 m^3/s after 15 days without rain, and 70 m^3/s after 45 days without rain. Derive the equation of the depletion curve in the form $Q_t = Q_o * e^{-at}$ and estimate the discharge 90 days without rain (B.Sc., DU, 2012).

Solution

1. Data: values of flow rate Q_a and Q_t, after one month and eight days, $Q_{15} = 140$, $Q_{45} = 70$ m^3/s. Q_{90}?
2. Find base flow hydrograph equation using the equation: $Q_t = Q_o * e^{-at}$

   Insert given values in the equation as shown below in equations 1 and 2.

   $140 = Q_o \times e^{-15a}$ (1)
   $70 = Q_o \times e^{-45a}$ (2)

   Dividing equations 1 and 2 to find values of coefficient : a = 0.0231 /day
3. Substitute in one of the equations 1 or 2 to find the value of the primary discharge Q_o as follows:

$Q_o = 140 \div$ e-15x0.0231 = 198 m^3/s

   Basal flow hydrograph formula becomes: $Q_t = 198 * e^{-0.0231t}$
4. Find rate of the stream after a period of 90 days in the watercourse compensation in the basal flow hydrograph equation obtained in step 4 above: $Q_t = 198 * e^{-0.0231 * t}$, $Q_{90} = 198 * e^{-90 \times 0.0231} = 24.7$ m^3/s

Example 6.7 (see Program 6.1 Listing)

The static level of water table in an unconfined aquifer was 33.5m above the underlying impermeable stratum. A 150mm diameter well, penetrating the aquifer to its full depth, was pumped at the rate of 25 litres per second. After several weeks of pumping, the drawdown in observation wells 20m and 50m from the well were 3.55m and

130

2.27m respectively, and the observed drawdowns were increasing very slowly. (B.Sc., UAE, 1989).

    a) Assuming equilibrium conditions, estimate the hydraulic conductivity and transmissivity of the aquifer.

    b) Estimate the drawdown just outside the pumped well.

    c) What will be the yield of a 300mm diameter well which will produce the same drawdowns just outside the well and at the 50m distance observation well in (ii)? What would be the drawdown at a nearer observation well?

Solution

i]

$$Q = \frac{\pi k (H^2 - h^2)}{\ln \dfrac{R}{r}}$$

$$\left[\frac{\pi k (H^2 - h^2)}{\ln \dfrac{R}{r}} \right]_{1\,st\,well} = \left[\frac{\pi k (H^2 - h^2)}{\ln \dfrac{R}{r}} \right]_{2\,nd\,well}$$

$$\frac{33.5^2 - (33.5 - 3.55)^2}{\ln \dfrac{R}{20}} = \frac{33.5^2 - (33.5 - 2.27)^2}{\ln \dfrac{R}{50}}$$

$$225.2475 \ln \frac{R}{50} = 146.9371 \ln \frac{R}{20}$$

$$\ln \frac{R}{50} = 0.6523 \ln \frac{R}{20}$$

$$\frac{R}{50} = \left(\frac{R}{20} \right)^{0.6523}$$

$$R^{0.34766} = 50 \left(\frac{R}{20} \right)^{0.6523}$$

∴ R = 279 m

$$k = \frac{Q. \ln \frac{R}{r}}{\pi(H^2 - h^2)} = \frac{20 \times 10^{-3} \ln \frac{279}{20}}{\pi[33.5^2 - (33.5 - 3.55)^2]} = \frac{20 \times 10^{-3} \ln \frac{279}{20}}{\pi[225 - (33.5 - 3.55)^2]}$$

$= 7.45 \times 10^{-5}$ m/s $= 6.44$m/d

T = kH = 6.44×33.5 = 215.7 m²/d

ii] Depth of water in well,

$$h^2 = H^2 - \frac{Q}{\pi k} \ln \frac{R}{r} = 33.5^2 - \frac{20 \times 10^{-3}}{\pi \times 7.45 \times 10^{-5}} \ln \frac{279}{0.075}$$

$= 419.7$

H = 20.49 m

Drawdown = 33.5 – 20.49 = 13.01 m.

iii]

a)

$$Q = \frac{\pi k (H^2 - h^2)}{\ln \frac{R}{r}} = \frac{\pi \times 7.45 \times 10^{-5}[33.5^2 - (33.5 - 2.27)^2]}{\ln \frac{279}{0.15}}$$

$= 4.6 \times 10^{-3} \, m^3/s$

b)

$$h^2 = 33.5^2 - \frac{4.6 \times 10^{-3}}{\pi \times 7.45 \times 10^{-5}} \ln \frac{279}{0.15} = 974.3$$

h = 31.2 m.

Drawdown = 33.5 – 31.2 =2.3m

Example 6.8 (see Program 6.1 Listing)

A well of diameter 0.3 m contains water to a depth of 50 m before pumping commences. After completion of pumping the draw-down in a well 20 m away is found to be 5 m, while the draw-down in another well 40 m further away reached 3 m. For a pumping rate of 2500 L/minute, determine:

- radius of zero draw-down.
- coefficient of permeability, and
- draw-down in the pumped well (B.Sc., DU, 2013).

Using expected well yield estimates as presented in table (5.1), comment about yield of the well as related to its diameter, & suggest a more suitable well diameter. Explain and validate your answer(B.Sc., DU, 2013) .

$$Q_o = \frac{\pi k \left(H^2 - h_o^2 \right)}{Ln \dfrac{R}{r_o}}$$

Solution

- Given: D= 0.3 m, H = 50 m, r_1 = 20 m, x_1 = 5 m, r_2= 40 m, x_2= 3 m, Q_o = 2500 L/min.

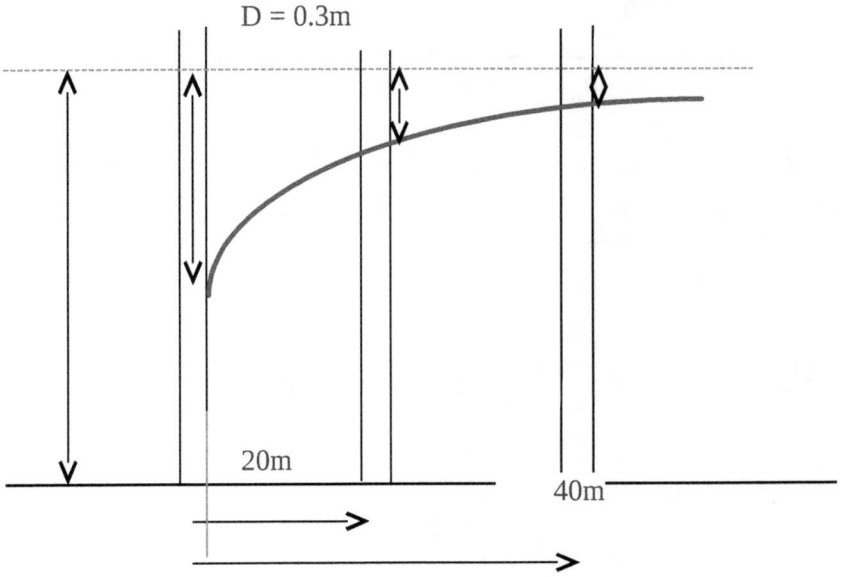

D = 0.3m

20m

40m

Find $h_1 = h - x_1 = 50 - 5 = 45$ m, and $h_2 = h - x_2 = 50 - 3 = 47$ m.
Use the following equation for both observation wells:

$$Q_o = \frac{\pi k \left(H^2 - h_0^2 \right)}{Ln \dfrac{R}{r_o}}$$

$$\therefore \left[\frac{\pi k \left(H^2 - h_0^2 \right)}{Ln \dfrac{R}{r_o}} \right]_{ist\ well} = \left[\frac{\pi k \left(H^2 - h_0^2 \right)}{Ln \dfrac{R}{r_o}} \right]_{2nd\ well}$$

By substituting given values into the previous equation, then:

$$\left[\frac{\left(15^2 - 13.7^2 \right)}{Ln \dfrac{R}{40}} \right]_1 = \left[\frac{\left(15^2 - 14.3^2 \right)}{Ln \dfrac{R}{90}} \right]_2$$

- This yields R = 119.7 m.
- Find the permeability coefficient by using the data of one of the wells.

   Thus, for h = 50 m, ho = 45 m, r = 20 m, R = 119.7 m, Q = $2500*10^{-3}*60*24 = 3600$ m³/day,

$$k = \frac{Q \ln \dfrac{R}{r_1}}{\pi \left[H^2 - h_o^2 \right]} = \frac{3600 \ln \dfrac{119.7}{20}}{\pi \left[50^2 - 45_1^2 \right]} = 4.32\,m/d$$

- Depth of the water in the pumped well may be found as:

$$Q_o = \frac{\pi k \left(H^2 - h_o^2 \right)}{Ln \dfrac{R}{r_o}} \qquad \text{Or}$$

$$h_0^2 = H^2 - \frac{Q}{\pi k} \ln \frac{R}{r_0} = 50^2 - \frac{3600}{\pi x\, 4.32} \ln \frac{119.7}{0.15}$$

This yields, h = 27 m

   Determine the draw-down at the well as:

   $r_o = h - h_1 = 50 - 27 = 23$ m.

- From table for a well diameter of 30 cm, yield is 2000 – 3500 m³/d.

  The amount of water abstracted is Q_o = 2500 L/min = 2500*60*24/1000 = 3600 m³/d. A better design would be selected a well of diameter of 35 cm (giving a yield of 3683 m³/d for computed drawdown sat well (O.k. between 3000 - 5000).

$$Q_o = \frac{\pi \times 4.32\left(50^2 - 27^2\right)}{Ln\dfrac{119.7}{(0.35/2)}} = 3683 \text{ cubic meter per day}$$

Example 6.9

1) A well penetrates into an unconfined aquifer having a saturated depth of 100 m. The discharge is 250 litres per minute at 12 m drawdown. Assuming equilibrium flow conditions and a homogeneous aquifer, estimate the discharge at 18 m drawdown. The distance from the well where the drawdown influence are not appreciable may be taken to be equal for both cases. (UAE, 1990).

Solution:

a) – Pollutants.   - Difficulty.

$$Q = \frac{\pi k\left(H^2 - h_1^2\right)}{\ln\dfrac{R}{r_o}} \qquad Q_1 = 250 = \frac{\pi k\left[100^2 - (100-12)^2\right]}{\ln\dfrac{R}{r_o}}$$

b) $\dfrac{\pi k}{\ln\dfrac{R}{r_o}} = \dfrac{250}{(100+88)(100-88)} = \dfrac{250}{188 \times 12}$

$$Q_2 = \frac{\pi k}{\ln\dfrac{R}{r_o}}[100^2 - (100-12)^2] = \frac{250}{188 \times 12}[100^2 - 82^2]$$

$$\frac{250}{188 \times 12} \times 182 \times 18 = 363 \, L/min$$

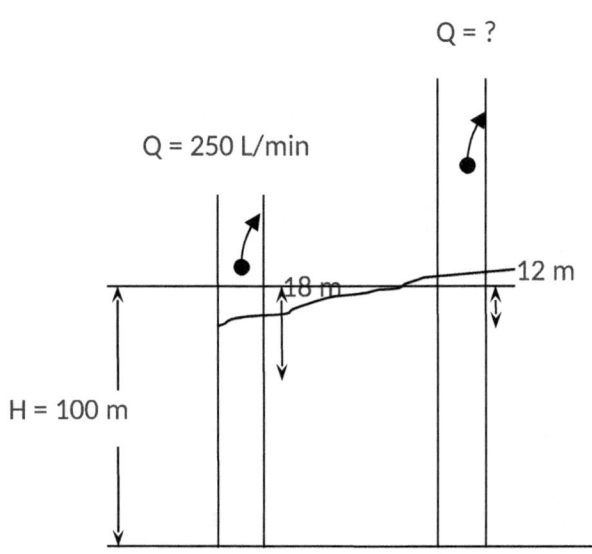

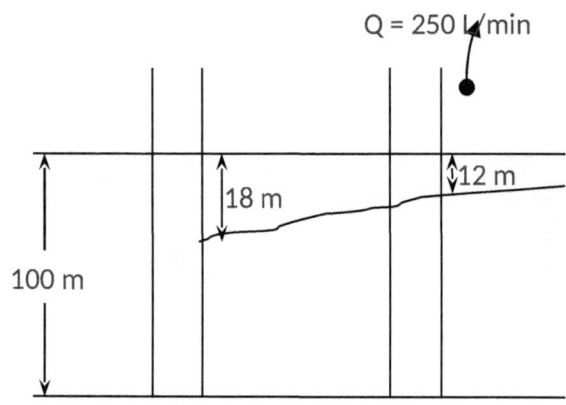

Example 6.10

A 0.3 m diameter well penetrates vertically through an aquifer to an impervious strata which is located 18 m below the static water table. After a long period of pumping at a rate of 1 m³/min., the drawdown in test holes 14 and 40 m from the pumped well is found to be 2.62 and 1.5 m, respectively.
1) Determine coefficient of permeability of the aquifer.
2) What is the transmissibility of the aquifer?
3) Compute the specific capacity of the pumped well. (UAE, 1990).

Solution:

1. r, Q, H, T, specific yield storage, lateral extent of aquifer, well

$$Q=\frac{\pi k \left(H^2-h^2\right)}{\ln \dfrac{R}{r}}$$

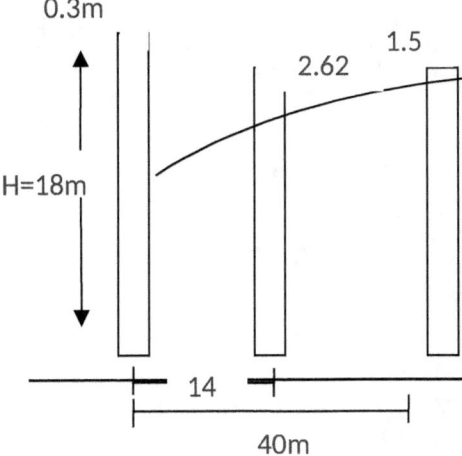

2.

$$\left[\dfrac{\pi\,k\left(18^2-\left[18-2.62\right]^2\right)}{\ln\dfrac{R}{14}}\right]_{(first\;well)}=\left[\dfrac{\pi\,k\left(18^2-\left[18-1.5\right]^2\right)}{\ln\dfrac{R}{40}}\right]$$

$$\ln\dfrac{R}{40}=0.5917\ln\dfrac{R}{14}$$

$$R^{0.40627}=\dfrac{40}{14^{0.5917}}=8.3926\rightarrow R=183.21\,m$$

i]

$$k=\dfrac{Q\ln\dfrac{R}{r}}{\pi\left(H^2-h^2\right)}(first\;well)=\dfrac{\dfrac{1}{60}\ln\dfrac{188.21}{14}}{\pi\left(18^2-15.38^2\right)}=1.56\,x\,10^{-4}\,m/s$$
$$= 13.5 \text{ m/d}$$

ii] T = kH = 13.5x18 = 243 m³/m.d

iii] depth in well

$$h^2=H^2-\dfrac{G}{\pi k}\ln\dfrac{R}{r}=18^2-\dfrac{1}{60\pi\,x\,1.56\,x\,10^{-4}}\ln\dfrac{183.21}{0.15}=82.28$$

h = 9.1 m

drawdown (r) = H −h = 18-9.1 = 8.9m

iv] specific capacity = $\dfrac{G}{r}=\dfrac{1}{60\,x\,8.9}m^2/s$

6.2 Theoretical Exercises

1. Outline main differences between confined and unconfined aquifer (B.Sc., DU, 2012).
2. Define the term "**Transmissibility** of aquifer".
3. Define parameters that may influence **yield** of wells.
4. Comment on yield of well compared to its diameter. Give suggestions for improvement?
5. Define factor that may influence yield of wells. (B.Sc., UAE 1989,, UAE 1990, B.Sc., DU, 2012).

Solution

Drawdown, Flow conditions, Depth of penetration, Transmissivity, Specific yield, Storage coefficient, Lateral extent of aquifer-, Construction and condition of well.

6. Outline most important factors affecting productivity of a well (B.Sc., DU, 2013).

7. In your opinion, what is the most significant source of groundwater pollution in this country? Explain why is groundwater contamination so difficult to detect and clean up? (UAE, 1990).

8. Define parameters shown in Theims equation: (B.Sc., DU, 2012)

$$S_o = \frac{Q_o}{2\pi kH} \ln \frac{R_o}{r} = \frac{Q_o}{2\pi T} \ln \frac{R_o}{r}$$

9. Define terms used in the equation used to estimate rate of constant pumping from a well penetrating an unconfined aquifer (B.Sc., DU, 2011)

$$Q_o = \frac{\pi k \left(H^2 - h_o^2 \right)}{Ln \dfrac{R}{r_o}}$$

10. Comment about Theim's equation (equilibrium equation) assumptions (B.Sc., DU, 2012).

11. Write briefly about groundwater in the kingdom of Saudi Arabia with reference to the following: Occurrence, Distribution, Quantity & volumes of aquifers, Quality, Potential hazards & pollution indicators.

12. What are the factors affecting groundwater flow within a basin?

13. What are Dubois's assumptions? How to use them?

14. What is the difference between stable and unstable flow?

15. What is the difference between a water table well and an artesian well? (Draw sketches).

16. Compare between confined, unconfined and perched aquifers. Illustrate your answer with suitable sketches. (B.Sc., UoD 2013, 2014)

6.3 Problem solving in groundwater
Well yield, drawdown

1. A 30 cm well is pumped at the rate of Q m^3/minute. At observation wells 15 m and 30 m away the drawdowns noted are 75 and 60 cm, respectively. The average thickness of the aquifer at the observation wells is 60 m and the coefficient of its permeability amounts to 26 m/day.
    1) Find the coefficient of **transmissibility** of the aquifer?
    2) Determine the **rate of pumping** Q.
    3) Compute **specific capacity** of the well.
    4) What is the **drawdown** in the pumped well? (UAE, 1989).
        (Ans. 65 m^2/hr, 1.5 m^3/min, 0.73 m^2/min, 2 m).

2. A 30 cm well penetrates 45 m below the static water table. After a long period of pumping at a rate of 1200 Lpm, the drawdown in the wells 20 and 45 m from the pumped well is found to be 3.8 and 2.4 m respectively.
    a) Determine the **transmissibility** of the aquifer.
    b) What is the **drawdown** in the pumped well? (UAE, 1989).
        (Ans. 7.1 m^2/hr, 13.5 m).

3. A 30 cm well serves a community of 1100 capita with a water consumption rate of 250 l/day. Under steady-state conditions the drawdown in the well was 1.83 m. The well penetrates to an impermeable stratum 32 m below the water table of a homogeneous-isotropic unconfined aquifer. Compute the change in the **discharge** in l/s for a well drawdown of 1.83 m if the diameter of the well was: a) 20.0 cm; and b) 50 cm. Assume that the radius of influence in all cases is 760 m. (SQU, 1991).

4. A community of 15,000 capita and fire demand of 35 L/s for six hours is to be served by a 50 cm diameter well. The well is constructed in a confined aquifer with a uniform thickness of 15 m and hydraulic conductivity of 100 m/d. Two observation wells are installed at radial distance of 50 m and 150 m. The drawdowns in the wells are 1.7 m and 1.3 m respectively. Find:
    a) the **discharge** of the well.
    b) water **consumption** (l/c/d).
    c) the **power** needed to lift the water to the ground surface if the original piezometer level is 20 m below the ground surface. (SQU, 1991).

5. Gravity well is 50 cm in diameter. The depth of water in the well is 30 meters before the start of pumping. Upon pumping at a rate of 2,100 liters per minute, Water level into a well located about 10 meters away fell by 3 meters, and in another well 20 meters away it reached 1.5 meters. Find: the zero draw down, permeability coefficient, and the drawdown in the well. Draw cone of depression attributed to pumping of this well. (Ans. 41.5m, 3024 m³/d, 5.2m).

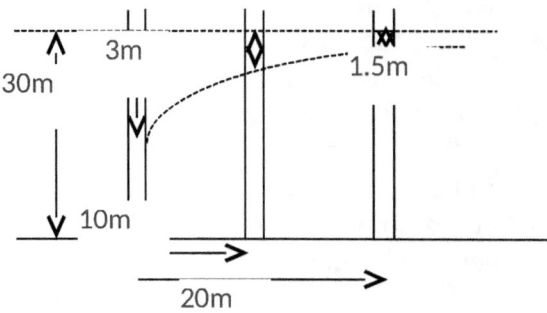

6. The static level of the surface of the water in an unconfined aquifer is 33.5 meters above the underneath impervious layer. Pumping is conducted at a rate of 20 liters per second in a 150 mm diameter well penetrating through all the depth of the aquifer. After a few weeks of pumping in observation wells at 20 meters and 50 meters distance the water table decreased by 3.55 meters and 2.27 meters, respectively and drawdowns increased gradually and slowly. Assuming equilibrium conditions, find: hydraulic permeability coefficient, drawdown directly outside the pumped well. Draw cone of depression attributed to pumping of this well. How much is the yield from a 300 mm diameter well that can achieve the same reduction in level outside it and at 50 meters from the observation well. Determine the amount of drawdown in the neighboring observation wells?

7. A well of diameter of 0.3 meters was drilled to impervious base in the center of a circular island of radius of one kilometer in a large lake. The well penetrates deeply in a sandstone aquifer which have a thickness of 20 meters which lies below a layer of

impermeable clay. Permeability of sandstone equals 20 meters per day. Find steady flow on the assumption that drawdown for the pezometic surface is equal to 2.5 meters in the well. (Assume R_0 = twice distance between aquifer and lake). Comment about yield of well as related to its diameter, and suggest a more suitable well diameter. Explain your answer.

8. A fully penetrating well, with an outside diameter of 0.2 m discharges water at a constant rate of 6 cubic meters per minute from an aquifer whose coefficient of transmissivility is 120 square meters per hour. The aquifer is in contact with a lake 1.6 kilometer away and has no other source of supply. What is drawdown at the well surface? (R_0 can be estimated to equal twice the distance between the aquifer and the lake).

9. Tube well 30 cm diameter penetrates an unconfined aquifer. Find the discharge from the tube well on the assumption that the drawdown does not exceed 2 meters, the effective height of the screen under conditions of the above mentioned drawdown is equal to 10 meters, the coefficient of permeability of the aquifer of 0.05 cm/s, and the radius of zero drawdown Ro (effective radius of drawdown) is equivalent to 300 meters.

10. A gravity well of diameter 30 cm, water depth in it 20 meters before pumping is started. When pumping is being done at a rate of 1800 liters per minute, the drawdown in a well 15 meters away is 2.5 m; and in another well 30 meters away is 1.5 m. Find:
   • Radius of zero draw down.
   • The coefficient of permeability, and
   • Drawdown in the well.
   • Give a suggestion for type of aquifer soil with compute permeability.

11. Find amount of water from a well of a diameter of 15 cm penetrating to the end of an unconfined groundwater aquifer at a distance of 30 m from the ground surface, note that its permeability is 3 meters/hour, and the level of drawdown in the well is 2.5 m, and radius at which the level of groundwater vanishes is 400 meters.

12. A well of diameter 30 cm penetrates into groundwater aquifer. Water depth in it reaches 15 meters. The level of fall of water in

two observation wells at distances 30 and 60 meters from the well was found to be 1.2 and 0.5 meters, respectively, when the water pumping rate is 4000 liters per minute. Find the permeability coefficient of the basin and the level of drawdown in the well after pumping.

13. An aquifer is unconfined, homogeneous, isotropic, of infinite areal extent, and of uniform thickness over the area influenced by the test. Prior to pumping, the piezometric surface is horizontal over the area that will be influenced by the test. The aquifer is pumped at a constant discharge rate. A well of diameter 30 cm, penetrates the entire saturated thickness of the aquifer. The gradient between the pumping well and monitoring wells is at steady-state. The velocity of flow is proportional to the tangent of the hydraulic gradient instead of the sine as it is in reality. The flow is horizontal and uniform everywhere in a vertical section through the axis of the well. Water depth in the well is 20 meters before pumping is started. When pumping is being done at a rate of 1800 liters per minute, the drawdown in a well 15 meters away is 2.5 m; and in another well 30 meters away is 1.5 m. Using Dupuit equation for steady-state Flow in the unconfined aquifer find: (B.Sc., UoD 2014)
    a) Radius of zero draw-down.
    b) The coefficient of permeability, and
    c) Drawdown in the well.
    d) Give a suggestion for type of aquifer soil with compute permeability.
    (Ans. 91.2 m, 15.9 m/d, 11.8 m)

$$Q = \frac{\pi k \left(h_2^2 - h_1^2 \right)}{\ln \dfrac{r_2}{r_1}}$$

Aquifer recharge

14. A well was drilled an impervious base in center of a circular island of diameter 1.5 km in a large lake. The well penetrates totally in the sandstone aquifer which is 18 meters thick located under a layer of impervious clay. Permeability of sandstone equals 15 meters per day. Find stable flow if the decline in

pesimoetric surface should not exceed 2 meters in a well of diameter 0.3 meters.

15. Suppose there are two canals, at different levels, separated by a strip of land 900 meter wide, of permeability that reaches 0.3 m/hr. If one of the canals is 1.3 m higher than the other and the depth of the aquifer is 15 m below the lower canal to an impermeable base, find inflow into, or abstraction from, each canal per unit width. take annual rainfall as 2.1 m/annum and assume 85 percent infiltration.

16. A rain gauge in a particular area showed that the average annual rainfall was 610 mm. 75 percent of this rain seeps into the ground. There are two channels separated by a piece of land of width 1 kilometer and permeability 0.35 m/h, one of the canals is higher than the other by about 1.2 meters. Geological surveys in the region pointed out to the presence of a groundwater basin under the piece of land with an average depth of 10 meters. Find the rate of outflow from each channel to the aquifer.

Chapter Seven

Surface runoff

7.1 Runoff

Runoff is that part of the precipitation, as well as any other flow contributions, which appears in surface stream of either perennial or intermittent form. It is the flow collected from a drainage basin or water shed, and it appears at an outlet of the basin. Specifically, it is the flow, which is the stream flow unaffected by artificial diversions, storage or any works of man in or on the stream channel, or in the drainage basin or water shed.

Runoff divisions according to source of flow [1,2,15] include: surface flow (runoff), sub surface flow (interflow, sub surface storm flow, storm seepage) and groundwater flow (groundwater runoff).

Example 7.1

An unregulated river has monthly mean flows (in m^3/s) as presented in the table (B.Sc., DU, 2013).

Table: Monthly records of stream flow.

Month	Monthly mean flow, m^3/s
January	5.4
February	8.3
March	9.1
April	8.8
May	6.3
June	6.9
July	10.2

August	13.7
September	19.4
October	16.7
November	11.0
December	21.9

1. Allowing compensation water of 4.0 m^3/s and reservoir losses of 0.5 m^3/s, what storage capacity of reservoir is required to ensure that, on average, no water is spilled? Assume 30-day months.
2. What would the average net yield of the reservoir then be? (Hint: average net yield of reservoir = demand – compensation – losses).

Solution

- Data: regular consumption of 230 m^3/min, & data of monthly water flow.
- Find cumulative total flow as shown in the following table:
- Draw mass curve for data by plotting cumulative flow values as a variable with time.

Month (1)	Flow, m^3/s (2)	Flow, m^3/day (3) = col (2)*60*60*24*30	Accumulated flow in month, m^3 (4)
Jan	5.4	13996800	13996800
Feb	8.3	21513600	35510400
Mar	9.1	23587200	59097600
April	8.8	22809600	81907200
May	6.3	16329600	98236800
June	6.9	17884800	1.16E+08
July	10.2	26438400	1.43E+08
Aug	13.7	35510400	1.78E+08
Sept	19.4	50284800	2.28E+08
Oct	16.7	43286400	2.72E+08
Nov	11	28512000	3E+08
Dec	21.9	56764800	3.57E+08

- Determine monthly flow (column 3) by multiplying each value in column (2) by 60*60*24*30 to obtain flow in m^3 per month, taking a 30day month as assumed.
- Find accumulated flow as shown in column (4).
- Draw mass curve of reservoir.
- Draw demand line for no water spillage on curve.
- Find storage for no spillage as maximum ordinate between mass curve and demand = $66*10^6$ m^3 = 66 Mm^3.
- Determine demand = $356.9*10^6/12 = 29.74*10^6$ m^3/month = $29.74*10^6/(30*24*60*60) = 11.5$ m^3/s
- Compute average net yield of reservoir = demand − compensation − losses = $11.5 − 4 − 0.5 = 7$ m^3/s.

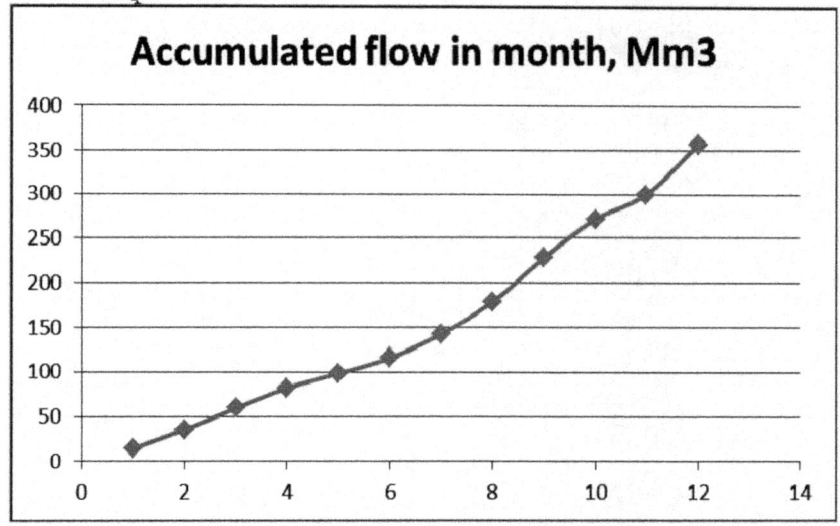

Program 7.1 Listing:

```
'*******************
'Example 7.1
'*******************
Public Class Form1
    Dim flow() As Double
    Dim comp, loss, demand, yield, N As Double
    Dim MAX_FLOW As Double

    Private Sub Form1_Load(ByVal sender As
        System.Object, ByVal e As System.EventArgs)
        Handles MyBase.Load
```

```vb
    Me.Text = "Example 7.1"
    Me.FormBorderStyle =
        Windows.Forms.FormBorderStyle.FixedDialog
    Label1.Text = "Enter monthly flow values:"
    Label2.Text = "Compensation (m3/s)"
    Label3.Text = "Reservoir loss (m3/s)"
    Label4.Text = ""
    Button1.Text = "&Calculate"
    DataGridView1.Columns.Clear()
    DataGridView1.Columns.Add("FlowCol",
        "Monthly mean flow (m3/s)")
    DataGridView1.Columns.Add("mFlowCol",
        "Flow (m3/day)")
    DataGridView1.Columns.Add("cFlowCol",
        "Monthly cumulative flow (m3)")
    DataGridView1.Columns(1).ReadOnly = True
    DataGridView1.Columns(2).ReadOnly = True
End Sub

Private Sub Button1_Click(ByVal sender As
    System.Object, ByVal e As System.EventArgs)
    Handles Button1.Click

    N = DataGridView1.RowCount - 1
    ReDim flow(N)
    comp = Val(TextBox1.Text)
    loss = Val(TextBox2.Text)

    Dim tmp As Double
    'calculate monthly flow for first month
    flow(0) =
        Val(DataGridView1.Rows(0).Cells(0).Value)
    flow(0) *= (60 * 60 * 24 * 30)

    'then start looping from the second month
    For i = 1 To N - 1
        tmp =
            Val(DataGridView1.Rows(i).Cells(0).Value)
        tmp *= (60 * 60 * 24 * 30)
        'display result on column (2) of the gridview
        DataGridView1.Rows(i).Cells(1).Value = tmp
        'calculate cumulative flow by adding
        'last month's flow
        flow(i) = tmp + flow(i - 1)
        'display result on column (3) of the gridview
        DataGridView1.Rows(i).Cells(2).Value = tmp
    Next

    MAX_FLOW = flow(N - 1)
```

148

```
    demand = flow(N - 1) / N
    demand /= (30 * 24 * 60 * 60)'convert to m3/s
    'compute net yield =
    '           demand - compensation - losses
    yield = demand - comp - loss
    Label4.Text = "Demand = " +
        FormatNumber(demand, 2) + " m3/s"
    Label4.Text += vbCrLf
    Label4.Text += "Net Yield = " +
        FormatNumber(yield, 2) + " m3/s"

    draw_graph()
End Sub

Private Sub draw_graph()
    Dim g As Graphics
    g = PictureBox1.CreateGraphics
    g.Clear(Color.White)

    Dim zX, zY As Double
    Dim scaleX, scaleY As Double
    'save (0,0) point
    zX = 5
    zY = PictureBox1.Height - 5
    'set scale factors for x & y axes
    scaleX = (PictureBox1.Width - 10) / N
    scaleY = (PictureBox1.Height - 10) / MAX_FLOW

    '***********************
    'draw x & y axes
    '***********************
    g.DrawLine(Pens.Black, CInt(zX), CInt(zY),
        CInt(zX), CInt(5))
    g.DrawLine(Pens.Black, CInt(zX), CInt(zY),
        CInt(PictureBox1.Width - 5), CInt(zY))
    'draw x-axis marks
    Dim f As Font = New
        Font(FontFamily.GenericMonospace, 8)
    For i = 1 To N - 1
        g.DrawString(i.ToString, f, Brushes.Black,
            CInt(zX + ((i - 1) * scaleX)),
            CInt(zY - 10))
    Next
    'draw y-axis marks at powers of 100, along
    'with their horizontal line mark
    For i = 1 To N - 1
        If (i * 100 * scaleY) > (PictureBox1.Height
                - 10) Then Exit For
        g.DrawString((i * 100).ToString, f,
            Brushes.Black, CInt(zX), CInt(zY -
```

```
            (i * 100 * scaleY)))
        g.DrawLine(Pens.Gray, CInt(zX), CInt(zY -
            (i * 100 * scaleY)),
            Cint(PictureBox1.Width - 5),
            CInt(zY - (i * 100 * scaleY)))
    Next
    '***********************
    'draw flow graph
    '***********************
    For i = 1 To N - 1
        g.DrawLine(Pens.Blue, CInt(zX +
            ((i - 1) * scaleX)), _
            CInt(zY - (flow(i - 1) * scaleY)), _
            CInt(zX + (i * scaleX)), _
            CInt(zY - (flow(i) * scaleY)))
        'draw a square mark
        g.DrawRectangle(Pens.Blue,
            CInt(zX + ((i - 1) * scaleX) - 2), _
            CInt(zY - (flow(i - 1) * scaleY) - 2),
            4, 4)
    Next
    'draw the last square mark
    g.DrawRectangle(Pens.Blue,
        CInt(zX + ((N - 1) * scaleX) - 2), _
        CInt(zY - (flow(N - 1) * scaleY) - 2), 4, 4)
    End Sub
End Class
```

7.2 Flow mass curve (Ripple diagram, S-curve)

This is a graph of the cumulative values of hydrological quantity (such as: runoff) plotted against time or data. Mass curve represents studying the effects of storage on the regime of a stream and of determining the regulated flow by means of a flow summation curve. Mass curve is a curve on which the ordinate of any point represents the total amount of water that has flowed past a given station on a stream during the length of time represented by the magnitude of the abscissa to the same point on the curve.

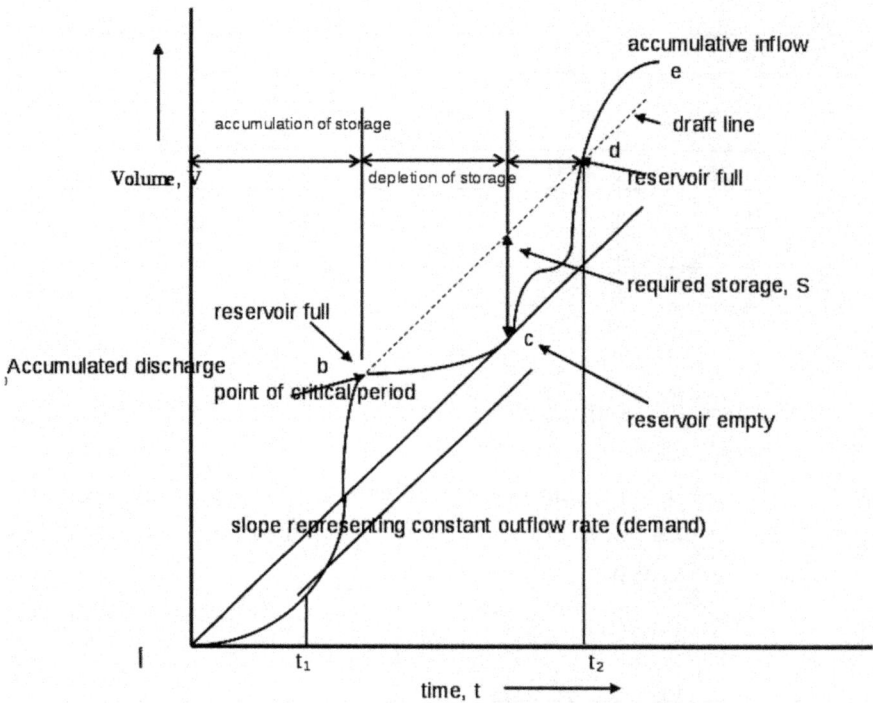

Fig. 7.7 Mass curve or Rippl diagram

Example 7.2

1. Outline benefits of a mass curve (flow duration curve or Rippl diagram).

2. A water reservoir is designed to collect water from the adjacent catchment basin & to regulate water use across an average regular flow of 230 cubic meters per minute. The table below shows the monthly records of the stream flow. Find amount of **storage** needed to keep up with regular consumption assuming no loss of water (B.Sc., DU, 2013).

Monthly records of stream flow.

Month	Volume of water, million cubic meter
January	8
February	65
March	45
April	35

May	25
June	12
July	2
August	3
September	9
October	45
November	67
December	77

Solution

- Data: regular consumption of 230 m^3/min, & data of monthly water flow.
- Find cumulative total flow as shown in the following table:
- Draw mass curve for data by plotting cumulative flow values as a variable with time.

Month	Volume of water (million cubic meter)	Cumulative Volume of water (million cubic meter)
1	8	8
2	65	73
3	45	118
4	35	153
5	25	178
6	12	190
7	2	192
8	3	195
9	9	204
10	45	249
11	67	316
12	77	393

- Find the value of the annual use rate (for the month of December) = 230 (m^3/min) × 60 (minutes/hour) × 24 (hours/day) × 365 (day/year) = 120.89 × 10^6 m^3/year = 121 million m^3/year.
- Draw draft line of uniform use from the point of origin to the point (a) on the mass curve.

152

- Draw a line parallel to the draft line from the point where the reservoir is full (b), & then find the value of minimum required storage for the reservoir to keep pace with consumption $= 20 \times 10^6 \text{ m}^3$.

Example 7.3

The peak water consumption on the day of maximum water usage as follows: (B.Sc., UAE, 1989).

Time	L/s	Time	L/s
Midnight	220	13	640
1	210	14	630
2	180	15	640
3	140	16	640
4	130	17	670
5	120	18	740
6	200	19	920
7	350	20	840
8	500	21	500
9	600	22	320
10	640	23	280
11	700	Midnight	220
Noon	660		

1. Calculate hourly cumulative consumption values.
2. Plot (i) to a mass diagram curve.
3. What is the constant 24 hour pumping rate.
4. Compute required storage capacity to equalize demand over the 24 hour period.

Solution

1. A graph of cumulative values of hydrologic quantities was plotted against time or data: Reservoir condition, Demand, Spillage

153

2.

Time	Hourly consumption		Comulative consumption
	L/s	L×10³	L×10⁶
Midnight	220	792	0.792
1	210	756	1.548
2	180	648	2.196
3	140	504	2.7
4	130	468	3.168
5	120	432	3.6
6	200	729	4.32
7	350	1260	5.58
8	500	1800	7.38
9	600	2160	9.54
10	640	2304	11.844
11	700	2520	14.364
12	660	2376	16.74
13	640	2304	19.044
14	630	2268	21.312
15	640	2304	23.616
16	640	2304	25.92
17	670	2412	28.32
18	740	2664	30.996
19	920	3312	34.308
20	840	3024	37.332
21	500	1800	39.132
22	320	1152	40.284
23	280	1008	41.292
Midnight	220	792	42.084

11590/24= 487

Average = 42084/(24*3600) = 487 L/s

7.3 Hydrograph

A hydrograph is a graph showing change of runoff volume (or stage, flow or discharge, velocity, or any other properties of water flow) w.r.t. time.

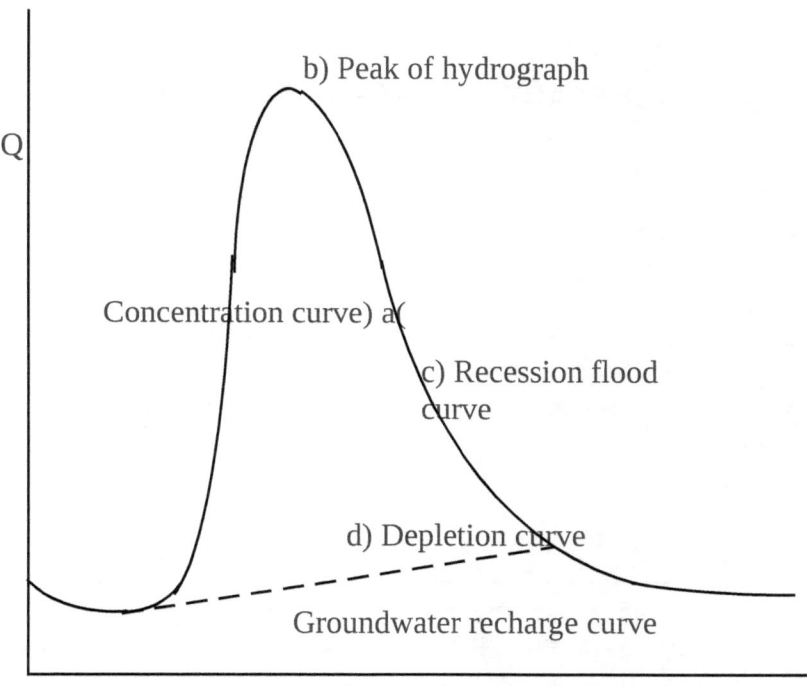

Fig. 7.2 Divisions of hydrograph

Example 7.4

The ordinates of discharge hydrograph measured on a 1-hour interval at Lahat Bridge. Pari river, Ipoh.

Ordinate Number	Accumulated Time	Discharge (m³/s)
1	9 pm	16.15
2	10 pm	13.22
3	11 pm	19.34
4	12 pm	28.00
5	1 am	16.55
6	2 am	15.11
7	3 am	14.05

8	4 am	13.00
9	5 am	11.00
10	6 am	11.59

CONSTANT DISCHARGE METHOD

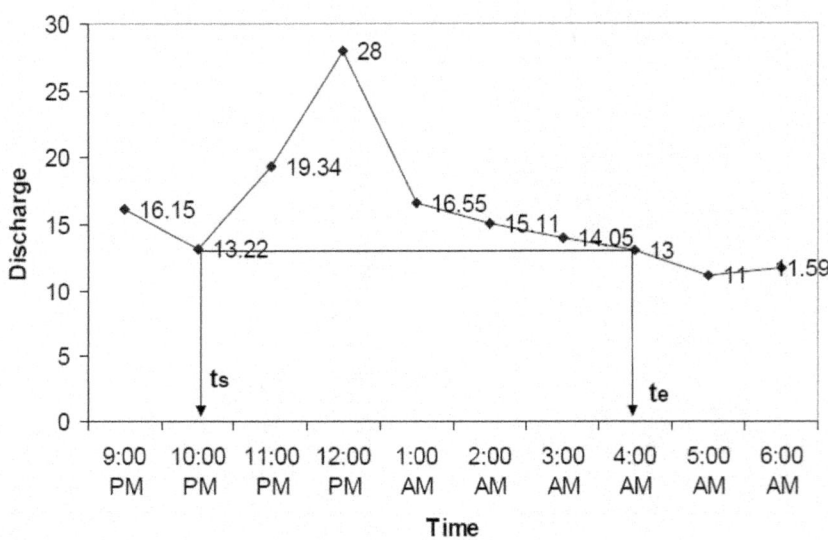

Constant Discharge Method

For t < ts $Qb = Q$

For ts ≤ t ≤ te $Qb = qs$

For t > te $Qb = Q$

Direct Runoff = Total Runoff - Base Flow

Ordinate Number	Accumulated Time	Discharge (m³/s)	Base flow (m³/s)	Direct Runoff (m³/s)
1	9 pm	16.15	16.15	0
2	10 pm	13.22	13.22	0
3	11 pm	19.34	10.29	9.05
4	12 pm	28.00	7.36	20.64
5	1 am	16.55	16.55	0

6	2 am	15.11	15.11	0
7	3 am	14.05	14.05	0
8	4 am	13.00	13.00	0
9	5 am	11.00	11.00	0
10	6 am	11.59	11.59	0

Equivalent Runoff Depth

Equivalent runoff depth = Total Direct Runoff X Time (s) + Area

Equivalent runoff depth = $(26.95 \times 60 \times 60)$ m^3 / (271×10^6) m^2 =

0.36×10^{-3} m = 0.36×10^{-3} m x 100 cm/m = 0.04 cm

Example 7.5

Solve the previous problem using the Concave Base Flow Method.

DISCHARGE HYDROGRAPH

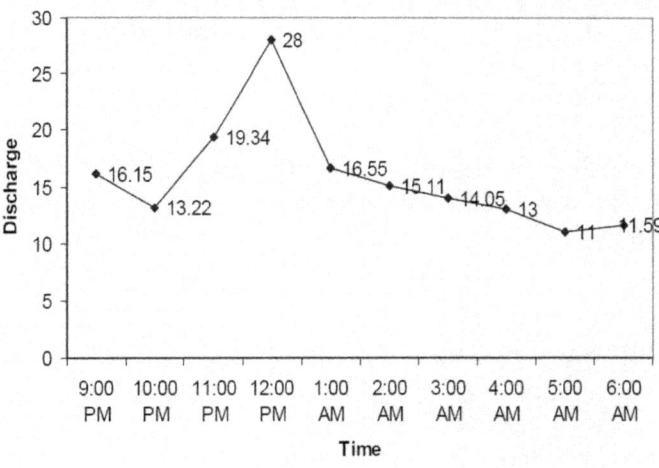

CONCAVE METHOD

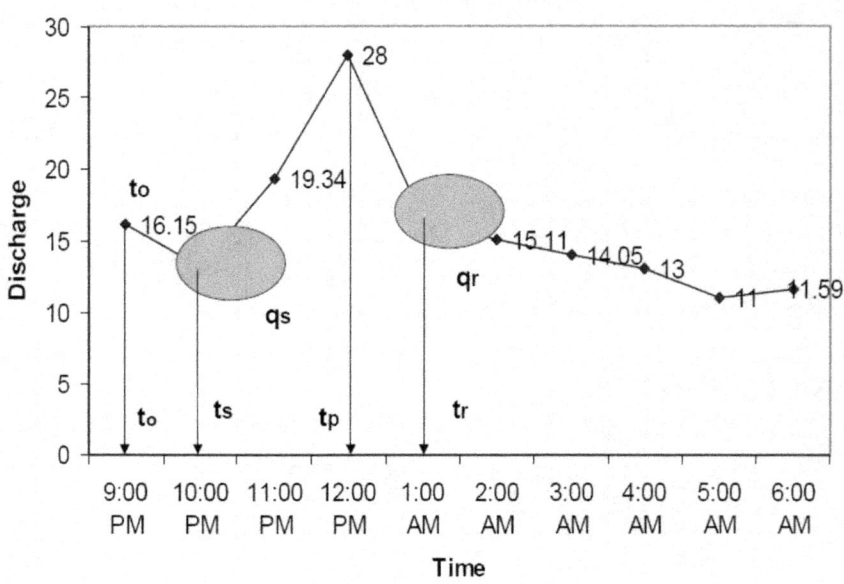

Important data taken from the hydrograph
T_s = 10.00 pm Q_s = 13.22 m³/s
T_r = 1.00 am Q_r = 16.55 m³/s
T_o = 9:00 pm Q_o = 16.15 m³/s
T_p = 12:00 pm
Notes: Q_b at tp is taken as q_m

$$q_b = \begin{cases} q \ for \ t < t_s \\ q_s + (t - t_s)[(q_s - q_o)/(t_s - t_o)] \ for \ t_s < t \le t_p \\ q_m + (t - t_p)[(q_r - q_m)/(t_r - t_p)] \ for \ t_p < t \le t_r \\ q \ for \ t_r < t \end{cases}$$

Example of calculation at 11:00 pm
Q_b = 13.22 + (11 – 10) [(13.22 – 16.15) / (10 – 9)] = 10.29 m³/s
Example of calculation at 1:00 am
Q_b = 7.36 + (1am – 12 am) [(16.55-7.36)/(1am – 12 am)]
      = 16.55 m³/s

Ordinate Number	Accumulated Time	Discharge (m³/s)	Base flow (m³/s)	Direct Runoff (m³/s)
1	9 pm	16.15	16.15	0
2	10 pm	13.22	13.22	0
3	11 pm	19.34	10.29	9.05
4	12 pm	28.00	7.36	20.64
5	1 am	16.55	16.55	0
6	2 am	15.11	15.11	0
7	3 am	14.05	14.05	0
8	4 am	13.00	13.00	0
9	5 am	11.00	11.00	0
10	6 am	11.59	11.59	0

Equivalent Runoff Depth
Equivalent runoff depth = Total Direct Runoff X Time (s) ÷ Area
Equivalent runoff depth = $(26.95 \times 60 \times 60)$ m³ / (271×10^6) m² = 0.36×10^{-3} m = 0.36×10^{-3} m x 100 cm/m = 0.04 cm
The general shape of the hydrograph of any ri

Example 7.6

A rain of 35mm fell on a drainage basin during a one-hour period beginning 01-00 hr. The drainage area is 78.2 km². The Following table lists the measured discharge from the area with respect to time: (B.Sc., UAE, 1989).

Hour	Discharge (m3/s)	Hour	Discharge (m3/s)
0	0	10	2.83
1	1	11	2.38
2	1.2	12	2.02
3	2.97	13	1.76
4	24.1	14	1.64
5	51.5	15	1.47
6	32.3	16	1.39
7	11.2	17	1.27
8	5.95	18	0
9	3.77		

1. Develop a one hour unit hydrograph.

2. Compute and plot the hydrograph of surface runoff for two periods of heavy rain occurring:

First storm: of 100mm rain between midnight and 02-00 hr; and 40mm rain between 02-00 and 03-00 hr.

Second storm: of 89mm rain between 04-00 and 05-00 hr. Assume a constant loss rate of 18mm and a constant base flow of 8m³/s

Comment about your graph.

Solution

1. A hydrograph of direct runoff resulting from 1mm of effective rainfall generated evenly over the basin area at a uniform rate.

   - Uniform effective rainfall.

   - Uniform distribution over area.

   - Constant time duration

   - Ordinate α Q

   - Reflects physical characteristics of basin.

2.
$$Average\ Q = \frac{140.75}{18} = 8.264\ m^3/s$$

Total runoff = $Q_a.t$ = 8.264×18 (hr)×3600 = 535500 m³

Measured discharge is to be divided by 6.85 to obtain UH

Time	m³/s	UH, m³/s/m	Time	82×UH	22×UH	71×UH	Base flow, m³/s	Total rainfall, m³/s
				Effective rainfall×UH				
0	0	0	24	0			8	8
1	1	0.15	1	12.3			8	20.3
2	1.2	0.18	2	14.76	0		8	22.76
3	2.97	0.43	3	35.26	3.3		8	43.56
4	24.1	3.52	4	288.64	3.96	0	8	300.6
5	51.5	7.5	5	615	9.46	10.65	8	643.11
6	32.3	4.7	6	385.4	77.44	12.78	8	483.95
7	11.2	1.6	7	131.2	165	30.53	8	334.73
8	5.95	0.87	8	71.34	103.4	249.92	8	432.66
9	3.77	0.55	9	45.1	35.2	532.5	8	620.8
10	2.83	0.41	10	33.62	19.41	333.7	8	394.44
11	2.38	0.35	11	28.7	12.1	113.6	8	162.4
12	2.02	0.29	12	23.78	9.02	61.77	8	102.57

13	1.76	0.26	13	21.32	7.7	39.05	8	76.07
14	1.64	0.24	14	19.68	6.38	29.11	8	63.17
15	1.47	0.21	15	17.22	5.72	24.85	8	55.79
16	1.39	0.2	16	16.4	5.28	20.59	8	50.27
17	1.27	0.19	17	15.58	4.62	18.46	8	46.66
18	0	0	18	0	4.4	17.04	8	29.44
			19		4.18	14.91	8	27.09
			20		0	14.2	8	22.2
			21			13.49	8	21.49
			22			0	8	8

Example 7.7

Draw the Unit Hydrograph for a river which flows in a midsize catchment of 120 km^2. After a 3-hour storm, the following flow hydrograph was recorded at the catchment outlet flow station. Assume a constant base flow of 10 m^3/s

Time (hr)	Flow Hydrograph (Q) (m^3/s)
0	10
3	20
6	35
9	50
12	70
15	50
18	35
21	10

Time (h)	Flow Hydrograph (Q) (m^3/s)	Base flow (m3/s)	Ordinates of DRH (m^3/s)		Ordinates of 3-hr UH (m^3/s)
0	10	10	0.0		0.0
3	20	10	10.0		5.0
6	35	10	25.0	/ ER	12.5
9	50	10	40.0		20.0
12	70	10	60.0		30.0
15	50	10	40.0		20.0
18	35	10	25.0		12.5
21	10	10	0.0		0

$\sum$ DRH = 200 m^3/s

Volume of Direct Runoff Hydrograph (DRH) = 60 X 60 X 3 X 200
= 2.16 Mm^3
Drainage Area = 120 km^2 = 120 X 10^6 m^2
Runoff Depth = Effective Rainfall (ER) = rainfall excess =
= Volume of DRH / Area = (2.16 X10^6) / (120 X10^6) 0.018 m = 1.8 cm
Approximately 2cm

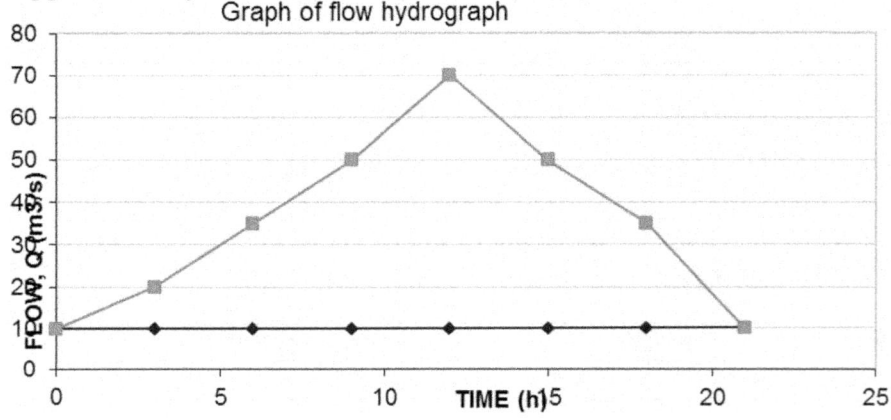

Check Area under UH = 1 cm over catchment area

Ordinates of 3-hr UH (m^3/s)
0.0
5.0
12.5
20.0
30.0
20.0
12.5
0

Σ DRH = 100 m^3/s

Volume = 60 x 60 x 3 x 100 km^2 = 1.08 x 10^6 m^3
Drainage area = = 120 km^2 = 120 x 10^6 m^2
Runoff Depth = Effective Rainfall (ER) = rainfall excess =
= Volume of DRH / Area = (1.08 x 10^6) / (120 X10^6) = 0.01m =
1.0 cm (OK)

162

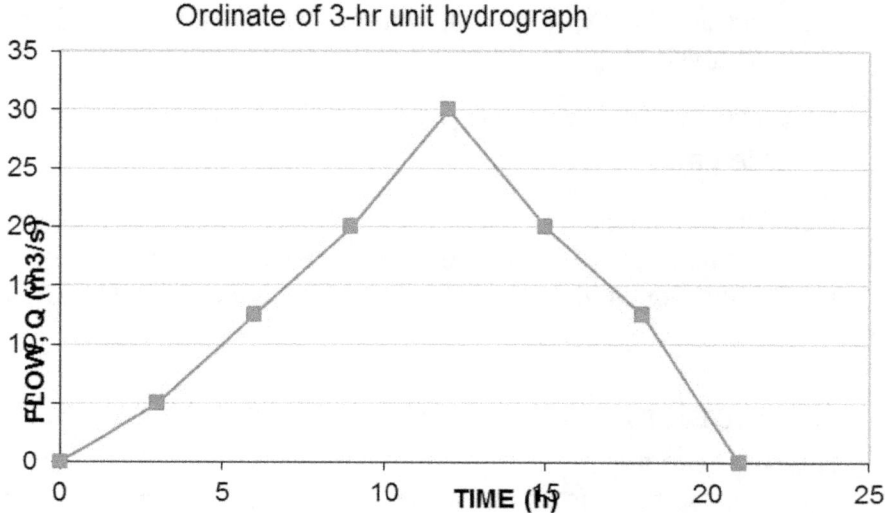

7.4 Theoretical Exercises

1. Write at length on each of the following: surface flow, subsurface and base flow, catchment area and use of radioactive materials.

2. Indicate ways used to measure surface flow? Which one would be preferred?

3. What are the assumptions included in the rational formula to estimate surface runoff?

4. Define mass curve "Rippl diagram". What information does it provide? (B.Sc., UAE, 1989).

5. Outline benefits of a mass curve (flow duration curve or Rippl diagram) (B.Sc., DU, 2013).

6. What is meant by "Unit Hydrograph"? Indicate assumptions involved in the hydrograph theory. (B.Sc., UAE, 1989).

7. Write briefly about each of: calibration curve, Rippl curve, and unit hydrograph.

8. How do you calculate peak flows upon availability and unavailability of measurements?

9. Write at length on each of the following: peak flows, and hydrograph.

10. What are the sections of a hydrograph? Indicate methods of its analysis?

11. What are the ways used to separate base flow in a hydrograph?

12. Write a detailed report on each of: unit hydrograph, instantaneous unit hydrograph, artificial water curve.

13. Outline potential uses of flow duration curves. (B.Sc., UoD 2014)

7.5 Problem solving in surface runoff
Rating curve

1) Suppose that the gauge shows a rise at the rate of 0.2 m/hr during a discharge measurement of 100 m³/s and the channel is such that this rate of rise may be assumed to apply to 1000 m reach of river between the measurement site and the reach control. Let the average width of the channel in the reach be 100 m, find the rate of change of storage, and discharge measurement to be plotted on the **rating curve.**

2) A river discharge measurement made during a flood indicated Qa = 2500 cubic meters per second. During the measurement, which took two hours, the gauge height increased from 30.2 meters to 30.4 meters. Level readings taken at water surface 340 meters upstream and 260 meters downstream of the observation site differed by 100 mm. the river was 350 meters wide with an average depth of 3 meters at the time of measurement. At what coordinate should the measurements be plotted on the **rating curve**?

3) A catchment area is undergoing a prolonged rainless period. The discharge of the stream draining it is 140 m³/s after 15 days without rain, and 70 m³/s after 45 days without rain. Derive the equation of the **depletion curve** in the form $Q_t = Q_o * e^{-at}$ and estimate the discharge 90 days without rain. (Ans. $Q_t = 198 * e^{-0.0231t}$,24.7 m³/s).

4) A catchment area is undergoing a prolonged rainless period. The discharge of the stream draining it is 100 m³/s after 10 days without rain, and 50 m³/s after 40 days without rain. Derive the equation of the **depletion curve** and estimate the

discharge 120 days without rain. (ans. $\alpha = 0.0231$ /day, $Q_t =$ $126.e^{-0.0231t}$, $Q_t = 7.9$ m^3/s)

5) A catchment area is undergoing a prolonged rainless period. The discharge of the stream draining it is 4100 m^3/min after ten days without rain, and 1200 m^3/min after one month without rain. Find:

   - The equation of **depletion curve**.
   - Estimate the discharge after a period of four months, and a period of six months without rain.
   - Draw depletion curve to scale.

6) Measuring the flow of a river during the flood has shown that Qa = 2700 cubic meters per second. During the measurement, which lasted for two hours, the level increased from 40.48 meters to 40.36 meters. Readings on the surface of the water level at distances of 390 meters and 310 meters upstream and downstream from the observation site differed by about 100 mm. the river width is 400 meters, and its average depth is 3.5 meters during the instant of measurement. Find coordinates of the point that should be taken for measurements in the **calibration curve**. (Ans. 40.42, 2639).

Storage

7) A water reservoir is designed to collect water from the adjacent catchment basin and to regulate water use across an average regular flow of 8000 cubic meters per hour. The following table shows the monthly records of the stream flow. Find amount of **storage** needed to keep up with regular consumption assuming no loss of water

Month	Volume of water (million cubic meter)	Month	Volume of water (million cubic meter)
January	7	February	21
March	15	April	14
May	10	June	5
July	1	August	1
September	4	October	14
November	26	December	32

8) A water reservoir is designed to collect water from the adjacent catchment basin and to regulate water use across an average regular flow of 250 cubic meters per minute. The following table shows the monthly records of the stream flow. Find amount of **storage** needed to keep up with regular consumption assuming no loss of water.

Month	Volume of water (million cubic meter)	Month	Volume of water (million cubic meter)
January	5	February	10
March	10	April	10
May	10	June	20
July	30	August	40
September	60	October	80
November	10	December	5

9) A reservoir of water is built to collect the amount of rain water falling in the neighboring basin and to organize a regular supply at a flow rate of 9000 cubic meters an hour. Records of stream flow indicated the monthly data listed in the following table. Find amount of **storage** required for organizing consumption (assuming no loss of water).

Month	Volume of water, million m^3
January	10
February	24
March	18
April	17
May	13
June	8
July	4
August	4
September	7
October	17
November	29
December	35

10) Define **mass curve** "Rippl diagram". What information does it provide? The peak water consumption on the day of maximum water usage as follows:

Time	L/s	Time	L/s
Midnight	220	13	640
1	210	14	630
2	180	15	640
3	140	16	640
4	130	17	670
5	120	18	740
6	200	19	920
7	350	20	840
8	500	21	500
9	600	22	320
10	640	23	280
11	700	Midnight	220
Noon	660		

- Calculate hourly cumulative consumption values.
- Plot (i) to a mass diagram **curve**.
- What is the constant 24 hour **pumping rate**.
- Compute required **storage** capacity to equalize demand over the 24 hour period. (ans. 487 L/s).

11) A reservoir is designed to collect water precipitated in the adjacent watershed basin adjacent to regulate supply with an average regular flow equal to 284 cubic meters per minute. The following table shows monthly capacity records of the stream in millions cubic meters. Find amount of **storage** needed to cope with regular consumption assuming no loss of water.

Month	volume of water (million cubic meters)	month	volume of water (million cubic meters)
January	7.4	February	49.4
March	31.1	April	34.6
May	29.6	June	8.6
July	2.5	August	1.2
September	18.5	October	58
November	76.5	December	82.7

12) The following is a record of the mean monthly discharge of a river in a dry year. Determine the minimum **capacity** of a water reservoir for a uniform draw-off of 50 cumsec, assuming no loss of water. (Take each month as 30 days for convenience) (B.Sc., UoD 2014)

Month	Mean inflow (cumsec)
January	35
February	80
March	72
April	62
May	28
June	31
July	73
August	55
September	80
October	72
November	45
December	31

Hydrograph

13) Flow rate in a stream discharging water shed runoff is 50 cubic meters per second after five days without rain. The flow rate in the stream reaches half that amount after 20 days without rain. Determine the equation of basal flow **hydrograph**. Find the amount of flow rate after a period of six months in the watercourse.

14) The rate of flow in a water course discharging a water shed is 3950 m³/min after ten days of no rain, and the flow rate is 1190 m³/min after one month with no rain. Draw the time **hydrograph** for base flow baseband; and find the amount of flow rate after three months and a five-month period in the watercourse. (Ans. 0.54, 0.015 m³/s)

15) Calculate and draw one-hour unit **hydrograph** for a certain area and noting that the discharge area is 70 square kilometers and the runoff from one peak for a precipitation of the 20 mm as represented in the following table:

Time, hr	Measured runoff, m^3/s
0	0
1	3.8
2	11.2
3	12.3
4	11.1
5	8.2
6	5.2
7	3.6
8	3.2
9	2.8
10	2.6
11	2.3
12	2.2
13	2.1
14	2.0
15	1.9
16	1.8
17	1.7
18	1.6
19	1.5
20	1.4
21	0

16) Calculate and draw one-hour unit **hydrograph** for to a certain area noting that 35 mm of rain fell in a discharge area of 78.2 square kilometers over a period of one hour started at 01:00. The following table shows the measured runoff through the area in the unit of time:

Time, hr	Measured flow, m^3/s	Time, hr	Measured flow, m^3
0	0	10	2.83
1	1	11	2.38
2	1.2	12	2.02
3	2.97	13	1.76
4	24.1	14	1.64

5	51.5	15	1.47
6	32.3	16	1.39
7	11.2	17	1.27
8	5.95	18	0
9	3.77		

Chapter Eight

Flow Routing

Flood means "any relatively high flow that overtops the natural or artificial banks in any reach of a stream".

Flood stage is the gauge height at which over flow of natural banks of a stream begins to cause damage. Floods may be measured as to height, area inundated, peak discharge & volume of flow. Flood elevation is measured by gauges (recording or non-recording type e.g. staff, post). Flood discharge can be measured by: direct methods (current meters), indirect ways: slope area, contractions, culverts, flow over dams & embankments, hydraulic methods, hydrologic data, & empirical methods.

Flood volume is measured by computation for a reservoir, daily discharge, & other means.

Flood routing is a procedure through which variation of discharge with time at a point on a stream channel can be determined by consideration of similar data for a point stream. Thus flood routing is a process that shows how a flood curve can be reduced in magnitude & lengthened in time (attenuated) by use of storage in the reach between two points. It is the process of determining progressively timing & shape of a flood wave at successive points along a stream.

Flood volume develops a procedure to determine the flow hydrograph at a point on a watershed from a known hydrograph upstream. As the hydrograph travels, it attenuates and gets delayed.

171

Evaluation of storage in a reach can be done by: making a detailed topographical & hydrological survey of the river reach & riparian land & so determine the storage capacity of the channel at different levels, or by using records of past levels of flood waves at the two points & hence deduce the reach's storage capacity.

Flood routing methods include:

- Invariable Discharge-Storage Relationship (The Puls method and the coefficient method).
- Variable Discharge-Storage Relationship (The Muskingum method).
- Lagging methods (The successive average-lag method and the progressive average-lag method).
- Graphic methods (The simplified Muskingum method and the working-value method).

In hydrologic Routing Input, output, and storage are related by continuity equation for unknown Q and S are as shown in equation 7.1

$$\frac{dS}{dt} = I(t) - Q(t)$$

(8.1)

Where:

I(t) = input

Q(t) = output

Storage can be expressed as a function of I(t) or Q(t) or both as depicted in equation 8.2.

$$S = f\left(I, \frac{dI}{dt}, \ldots, Q, \frac{dQ}{dt}, \ldots\right)$$

(8.2)

For a linear reservoir, S = kQ   (8.3)

Or as shown in equation 8.4.

$$k\frac{dQ}{dt} + Q(t) = I(t)$$

(8.4)

$$\Omega = \frac{Q(t)}{I(t)} = \frac{1}{1+kD}$$

(8.5)

172

Using Muskingum Method in hydrologic river routing, storage is linear function of I and Q that can be represented by equation 8.6.

$$S = K[XI + (1-X)Q]$$ (8.6)

Where

K = travel time of peak through the reach

X = weight on inflow versus outflow ($0 \le X \le 0.5$)

X = 0 for reservoir, storage depends on outflow, no wedge storage.

X = 0.0 - 0.3 for natural stream

And

$$S_{j+1} - S_j = K\{[XI_{j+1} + (1-X)Q_{j+1}] - [XI_j + (1-X)Q_j]\}$$ (8.7)

Recalling equation 7.8

$$S_{j+1} - S_j = \frac{I_{j+1} + I_j}{2}\Delta t - \frac{Q_{j+1} + Q_j}{2}\Delta t$$ (8.8)

Combining equation 8.7 and 8.8 yields equation 8.9.

$$Q_{j+1} = C_1 I_{j+1} + C_2 I_j + C_3 Q_j$$ (8.9)

Where:

$$C_1 = \frac{\Delta t - 2KX}{2K(1-X) + \Delta t}$$

$$C_2 = \frac{\Delta t + 2KX}{2K(1-X) + \Delta t}$$

$$C_3 = \frac{2K(1-X) - \Delta t}{2K(1-X) + \Delta t}$$

If I(t), K and X are known, Q(t) can be calculated using above equations

Q = Outflow

I = inflow

C1, C2, C3 = Muskingum constants

Dt = Routing period

K = travel time of peak through the reach = coefficient = storage constant

X = weight on inflow versus outflow ($0 \le X \le 0.5$)

Example 8.1

Given inflow hydrograph for a certain flow and values of K = 2.3 hr, X = 0.15, Δt = 1 hour, Initial Q = 92 cfs, determine Outflow hydrograph using Muskingum routing method.

Period, hr	Inflow, cfs
1	100
2	144
3	215
4	327
5	449
6	553
7	637
8	685
9	698
10	682
11	641
12	578
13	484
14	397
15	336
16	254
17	191
18	141
19	115
20	97

Solution

1.   Given: K = 2.3 hr, X = 0.15, Δt = 1 hour
2.   Using the given data find constants

$$C_1 = \frac{\Delta t - 2KX}{2K(1-X)+\Delta t} = \frac{1-2*2.3*0.15}{2*2.3(1-0.15)+1} = 0.0631$$

$$C_2 = \frac{\Delta t + 2KX}{2K(1-X)+\Delta t} = \frac{1+2*2.3*0.15}{2*2.3(1-0.15)+1} = 0.3442$$

$$C_3 = \frac{2K(1-X)-\Delta t}{2K(1-X)+\Delta t} = \frac{2*2.3*(1-0.15)-1}{2*2.3(1-0.15)+1} = 0.5927$$

Use Muskinum equation for routing flow:

$$Q_{j+1} = C_1 I_{j+1} + C_2 I_j + C_3 Q_j$$

Period, hr	Inflow, cfs	Q outflow, cfs
1	100	92
2	144	98
3	215	121
4	327	166
5	449	240
6	553	331
7	637	427
8	685	516
9	698	585
10	682	630
11	641	649
12	578	642
13	484	610
14	397	553
15	336	486
16	254	420
17	191	348
18	141	281
19	115	222
20	97	177

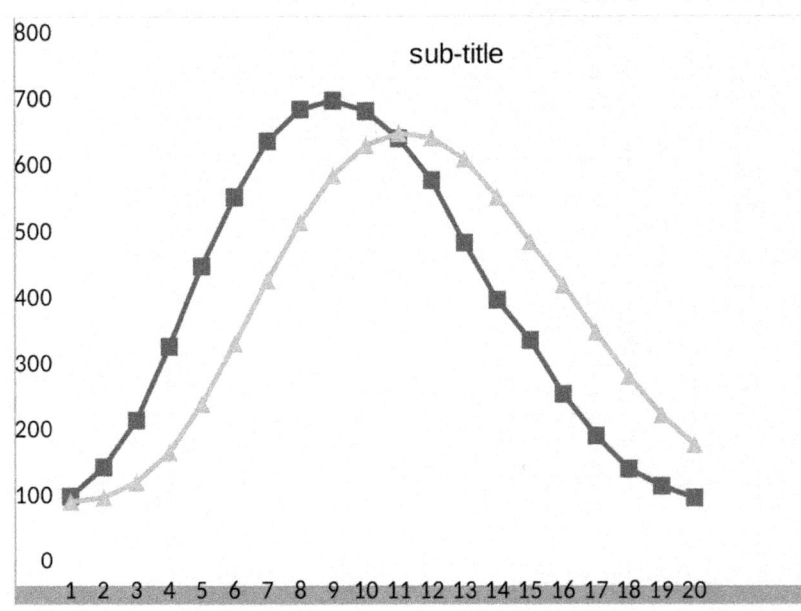

Program 8.1 Listing:

```
'*****************************************
'Example 8.1: Muskinum equation
'*****************************************
Public Class Form1
    Dim inflow(), outflow() As Double
    Dim K, X, Q, dt, N As Double
    Dim MAX_INFLOW, MAX_OUTFLOW As Double

    Private Sub Form1_Load(ByVal sender As
        System.Object, ByVal e As System.EventArgs)
        Handles MyBase.Load

        Me.Text = "Example 8.1: Muskinum equation"
        Me.FormBorderStyle =
            Windows.Forms.FormBorderStyle.FixedDialog
        Label1.Text = "Enter inflow hydrograph values:"
        Label2.Text = "K"
        Label3.Text = "X"
        Label4.Text = "dt"
        Label5.Text = "Init. Q"
        Button1.Text = "&Calculate"
        DataGridView1.Columns.Clear()
```

```vb
        DataGridView1.Columns.Add("inCol",
            "Inflow (cfs)")
        DataGridView1.Columns.Add("outCol",
            "Outflow (cfs)")
        DataGridView1.Columns(1).ReadOnly = True
    End Sub

    Private Sub Button1_Click(ByVal sender As
        System.Object, ByVal e As System.EventArgs)
        Handles Button1.Click

        N = DataGridView1.RowCount - 1
        ReDim inflow(N), outflow(N)
        Dim C1, C2, C3 As Double

        K = Val(TextBox1.Text)
        X = Val(TextBox2.Text)
        dt = Val(TextBox3.Text)
        Q = Val(TextBox4.Text)
        MAX_INFLOW = Q
        'save inflow data
        For i = 0 To N - 1
            inflow(i) =
        Val(DataGridView1.Rows(i).Cells(0).Value)
            If inflow(i) > MAX_INFLOW Then
                MAX_INFLOW = inflow(i)
        Next
        outflow(0) = Q
        DataGridView1.Rows(0).Cells(1).Value =
                outflow(0)
        C1 = (dt - (2 * K * X)) /
                ((2 * K * (1 - X)) + dt)
        C2 = (dt + (2 * K * X)) /
                ((2 * K * (1 - X)) + dt)
        C3 = ((2 * K * (1 - X)) - dt) /
                ((2 * K * (1 - X)) + dt)
        MAX_OUTFLOW = outflow(0)
        'calculate outflows
        For i = 1 To N - 1
            outflow(i) =
                (C1 * inflow(i)) +
                (C2 * inflow(i - 1)) +
                (C3 * outflow(i - 1))
            DataGridView1.Rows(i).Cells(1).Value =
                outflow(i)
            If outflow(i) > MAX_OUTFLOW Then
                MAX_OUTFLOW = outflow(i)
        Next
        draw_graph()
    End Sub
```

177

```vb
Private Sub draw_graph()
    Dim g As Graphics
    g = PictureBox1.CreateGraphics
    g.Clear(Color.White)

    Dim zX, zY As Double
    Dim scaleX, scaleY As Double
    'save (0,0) point
    zX = 5
    zY = PictureBox1.Height - 5
    'set scale factors for x & y axes
    scaleX = (PictureBox1.Width - 10) / N
    If MAX_INFLOW > MAX_OUTFLOW Then
        scaleY = (PictureBox1.Height - 10) /
            MAX_INFLOW
    Else
        scaleY = (PictureBox1.Height - 10) /
            MAX_OUTFLOW
    End If

    '***********************
    'draw x & y axes
    '***********************
    g.DrawLine(Pens.Black, CInt(zX), CInt(zY),
        CInt(zX), CInt(5))
    g.DrawLine(Pens.Black, CInt(zX), CInt(zY),
        CInt(PictureBox1.Width - 5), CInt(zY))
    'draw x-axis marks
    Dim f As Font = New
        Font(FontFamily.GenericMonospace, 8)
    For i = 1 To N - 1
        g.DrawString(i.ToString, f, Brushes.Black,
        CInt(zX + (i * scaleX)), CInt(zY + 1))
    Next

    'draw y-axis marks at powers of 100,
    'along with their horizontal line mark
    For i = 1 To N - 1
        If (i * 100 * scaleY) >
        (PictureBox1.Height - 10) Then
            Exit For
        g.DrawString((i * 100).ToString, f,
            Brushes.Black, CInt(zX),
            CInt(zY - (i * 100 * scaleY)))
        g.DrawLine(Pens.Gray, CInt(zX),
            CInt(zY - (i * 100 * scaleY)),
            CInt(PictureBox1.Width - 5),
            CInt(zY - (i * 100 * scaleY)))
    Next
```

178

```
'***********************
'draw inflow graph
'***********************
For i = 1 To N - 1
    g.DrawLine(Pens.Red, CInt(zX +
        ((i - 1) * scaleX)),
        CInt(zY - (inflow(i - 1) * scaleY)),
        CInt(zX + (i * scaleX)),
        CInt(zY - (inflow(i) * scaleY)))
    'draw a square mark
    g.DrawRectangle(Pens.Red, CInt(zX +
        ((i - 1) * scaleX) - 2),
        CInt(zY - (inflow(i - 1) * scaleY)
        - 2), 4, 4)
Next
'draw the last square mark
g.DrawRectangle(Pens.Red, CInt(zX +
    ((N - 1) * scaleX) – 2),
    CInt(zY - (inflow(N - 1) * scaleY)
    - 2), 4, 4)
'***********************
'draw outflow graph
'***********************
For i = 1 To N - 1
    g.DrawLine(Pens.Green, CInt(zX +
        ((i - 1) * scaleX)),
        CInt(zY - (outflow(i - 1) * scaleY)),
        CInt(zX + (i * scaleX)),
        CInt(zY - (outflow(i) * scaleY)))
    g.DrawRectangle(Pens.Green,
        CInt(zX + ((i - 1) * scaleX) – 2),
        CInt(zY - (outflow(i - 1) * scaleY)
        - 2), 4, 4)
Next
'draw the last square mark
g.DrawRectangle(Pens.Green, CInt(zX +
    ((N - 1) * scaleX) – 2),
    CInt(zY - (outflow(N - 1) * scaleY)
    - 2), 4, 4)
    End Sub
End Class
```

Chapter Nine

General exercises

9.1 Complete missing titles and fill in the blank spaces

1) **Complete missing titles by using the following words and phrases (Cyclonic precipitation. Relative humidity. Residence time. Unconfined aquifer. Depression storage. Speed. Detention storage)** (B.Sc., DU, 2012)

1. When rainfall is more than infiltration, evaporation and evapotranspiration, ponds gather in low-lying areas representing **depression storage**.
2. Filling of ponds and depressions Leads water to flood on the surface of the earth representing **detention storage**.
3. **Residence time** is the average duration for a water molecule to pass through a subsystem of the hydrologic cycle.
4. **Relative humidity** describes the ability of air to absorb additional moisture at a given temperature.
5. In absence of other factors tending to influence wind, it should be expected that its direction would be from areas of high pressure towards areas of low pressure and that its **speed** would vary with the pressure gradient.
6. **Cyclonic precipitation** is linked to passage over low temperature areas or altitude, resulting in lifting of hot air masses over cold masses.

7. Dupuit assumptions are used to get an approximate solution for uni-dimensional flow in an **unconfined aquifer**.

8. Fill-In-The Blanks using the Word Bank in the box: Every word can only be used once. (B.Sc., UoD 2014)

> Condensation. Cone of depression. Dew. Hydrologic cycle.
> Oceans. Permeability. Potable. Porosity. Unsaturated. Wind.

a. Potable water is good enough to drink
b. The ground water table is formed by the intersection of the zone of saturation with the unsaturated zone
c. The ability of a solid to transmit fluids is measured by its permeability.
d. Porosity is the amount of pore space in a rock unit.
e. A/an cone of depression forms as a result of excessive drawdown and the lowering of the water table around a well.
f. The oceans are great storehouses of water. They hold about 97% of the world's water.
g. The moisture-laden air is blown by the wind toward the land.
h. As air rises it cools. It eventually reaches the point where it can no longer hold all of its moisture in gaseous form. Some of the water vapor changes into a liquid around particles of dust, to form tiny droplets of water. When water vapor changes into water, the process is called condensation. We can see the results of this process in the atmosphere in the form of clouds that are carried by the wind.
i. Sometimes during cool evenings and early mornings, water vapour condenses as droplets on surfaces such as leaves, spider webs, cars and other cold surfaces. This moisture is called dew. Although this form of condensation does not remove much water from the atmosphere, it brings water to many life forms.
j. Much of the surface run-off and groundwater makes its way back to the oceans where it again may be evaporated into the atmosphere to start the hydrologic cycle all over again.

9.2 True/false sentences

2) **Indicate whether the following sentences are true (T) or false (F):** (B.Sc., DU, 2012)
   - The intensity and frequency of the hydrological cycle is independent of geography and climate. (**F**)
   - The role of applied hydrology is to provide guidance for planning and management of water resource. (**T**)
   - Earth's rotation affects movement of wind and rotation of seasons. (**T**)
   - Trees reduce wind speeds and cool the air as they lose moisture and reflect heat upwards from their leaves. (**T**)
   - Frost or ice is dew falling from the sky and freezes on the ground. (**T**)
   - Rate of evaporation decreases as the specific gravity increases. (**T**)
   - Infiltration devices estimate quality of infiltration rather than the quantity. (**T**)
   - A confined or artesian aquifer is one that is separated from the surface by an aquiclude or aquitard. (**T**)

3) **Indicate whether the following sentences are true (T) or false (F):** (B.Sc., DU, 2011)
   - Hydrology concerns quantity and quality of moving water accumulating on land, soil & rock adjacent to the surface. (**T**)
   - At night, treetops act as radiating surfaces, and the soil beneath is protected from excessive heat losses. (**T**)
   - In a large city, the amount of heat which is produced annually roughly equals the solar radiation reaching an equivalent area. (**F**)
   - Winds are mainly the result of horizontal differences in pressure. (**T**)
   - Thiessen method is suitable to find number of medium height to a particular area served by a network of fixed stations in number and positions. (**T**)
   - Method of isohyets maps give results close to those obtained by the Thiessen network method. (**T**)
   - Intensity of rainfall indicates quantity of rain falling in a given time. (**T**)

- Clouds slow up the process of evaporation. (**T**)
- Rate of evaporation is more for salt water than for fresh water. (**F**)
- The role of applied hydrology is to provide guidance for planning and management of water resource. (**T**)

9.3 Underline the best answer

4) **Underline the best word or phrase (between brackets) to offer a useful sentence.** (B.Sc., DU, 2011)

1) (**Residence time**/Time of concentration/Rainfall duration) Average travel time for water to pass through a subsystem of the hydrologic cycle.

2) (**Relative humidity** /Dew point/Humidity) is the ratio of amount of moisture in a given space to the amount a space could contain if saturated.

3) A (**psychrometer**/barometer/desiccant) is an instrument used to determine the value of humidity..

4) Actinometer and Radiometer are general names for instruments used to measure (**intensity**/duration/frequency) of radiant energy.

5) Winds are mainly the result of (vertical/**horizontal** /direct) differences in pressure..

6) An instrument for measuring the speed or force of the wind is called (wind vane/**anemometer**/pyranometer).

7) One of the most important reasons leading to condensation of vapor is (heating by air masses mix/ Dynamic or adiabatic heating/**Contact and radiational cooling**).

8) Moist air moving up the side of a mountain facing the prevailing wind causes precipitation to fall in a process known as(conventional/**orographic**/Cyclonic) lifting.

9) One of the factors that influence the measurement of rainfall, especially solid part, is (**wind**/pressure/latitude).

10) Of appropriate recommendations to determine the number of monthly rain gauge stations those proposed by (**Bleasdale**/Buys Ballot/Thiessen).

5) **Underline the best word or phrase (between brackets) to offer a useful sentence.** (B.Sc., DU, 2012)

11) (**Orographic precipitation___**/ Conventional precipitation / Cyclonic precipitation) occurs due to obstruction of topographical barriers (mountains, natural hills) for winds laden with moisture and its mechanical lifting to top layers and then its expansion & cooling resulting in rainfall.

12) (**Horton**/ Penman / Agot) equation shows a mathematical method to estimate the infiltration capacity curve.

13) (**Φ - index** / W-index / Antecedent Precipitation Index) is that rate of rainfall above which the rainfall volume equals the runoff volume.

14) (**Water-bearing layers** / Man-made reservoirs / Cones of depression) are geological features with permeability and they have components that allow significant movement of water through them.

15) (**Meteoric water** / Juvenile water /Magmatic water) denotes water in or recently from the atmosphere.

9.4 Rearrange groups

6) **Rearrange group (I) with the corresponding relative ones of group (II) in the area allocated for the answer in table (1).** (B.Sc., DU, 2012)

Table (1) Matching of relevant words or phrases..

Group (I)	Rearranged group (II)	Group (II)
Hydrology	**Water science**	Radiometer
Holton	**Infiltration**	Baric
Psychrometer	**Humidity**	Penman
Connate water	**Fossil water**	Unconfined
Condensation	**Contact cooling**	Water science
Perched aquifer	**Unconfined**	Limestone
Actinometer	**Radiometer**	Fossil water
Evapotraspiration	**Penman**	Infiltration
Karst	**Limestone**	Contact cooling
Buys-Ballot	**Baric**	Humidity

7) **Rearrange group (I) with the corresponding relative ones of group (II) in the area allocated for the answer.**
(B.Sc., DU, 2011)

Group (I)	Rearranged group (II)	Group (II)
Humidity	**Psychrometer**	Rainfall depth
Contact cooling	**Dew**	Convective precipitation
Acinometer	**Radiant energy**	Anemometer
Convectional precipitation	**Convective precipitation**	Hygrometer
Wind speed	**Anemometer**	Frontal precipitation
Wind direction	**Wind vane**	Dew
Orographic precipitation	**Windward slopes**	Radiant energy
Relative humidity	**Hygrometer**	Wind vane
Cyclonic precipitation	**Frontal precipitation**	Psychrometer
Rain gauge	**Rainfall depth**	Windward slopes

9.5 Complete missing titles

8) Complete missing titles by using the following words and phrases **(Dew point, Storage, Speed, Humidity, Temperature, Runoff, Actinometer, Water, Science, Wind)** (B.Sc., DU, 2011)

1. Hydrology is derived from the Greek words: hydro (meaning **water**) and logos (meaning dealing with **science**).

2. When rainfall is more than infiltration, evaporation and evapotranspiration, ponds gather in low-lying areas representing Depression **storage**.

3. The water cycle consists of precipitation, evaporation, evapotranspiration and **runoff.**

4. Earth's rotation affects movement of **wind** and rotation of seasons.

5. Relative **humidity** is the percentage of actual vapor pressure to saturation vapor pressure.

185

6. **Dew point** denotes the temperature at which space becomes saturated when air is cooled under constant pressure & with constant water vapor pressure.

7. Shading effect of trees tends to keep daily maximum **temperature** somewhat lower.

8. **Actinometer** and Radiometer are general names for instruments used to measure intensity of radiant energy..

9. **Speed** would vary with the pressure gradient.

9.6 Match the words or phrases

Adiabatic	Snow	Orographic	Precipitation	Convective
Rain gauge	Cyclone	Dew	Winds	Baric

9) Match the words or phrases above with the definitions below. (B.Sc., DU, 2011)

   a) Are mainly the result of horizontal differences in pressure.. (Winds)

   b) Baric wind law is also known as Buys-Ballot's law. (Baric)

   c) In this type of cooling heat is not added from outside sources. (Adiabatic)

   d) It forms when cloud droplets (or ice particles) in clouds grow and combine to become so large that the updrafts in the clouds can no longer support them, and they fall to the ground. (Precipitation)

   e) Represents frozen water pouring from clouds in small pieces. (Snow)

   f) Denotes water vapor that condenses in the cold layers of the atmosphere during the night and falls to the ground in small droplets. (Dew)

   g) In this type of precipitation most rain is deposited on the windward slopes. Other parts are located under the rain shadow and remain dry. (Orographic)

   h) This type of precipitation results from lifting of air converging into a low pressure area. (Cyclone)

   i) This type of precipitation is local, and its intensity varies from light rain showers to dangerous and destructive thunderstorms. (Convective)

j) It is a hydrological instrument used in measurements to calculate rainfall for the region. (Rain gauge)

9.7 General questions.

1.   This semester focused on hydrological cycle, water balance, meteorological factors, surface and ground water flow through landscape. For each of the following topics you are requested to identify the most important equation or principle studied. In each case, write the equation or define the principle and state briefly why you selected this equation or principle instead of alternatives that you might have chosen. (B.Sc., UoD 2013)

a) Hydrologic cycle, control volumes, hydrologic system

b) Energy balance, atmospheric circulation, water vapor flow

c) Precipitation formation and measurement

d) Evaporation

e) Infiltration and soil water movement

f) Storm runoff

g) Groundwater flow

h) Nonpoint source pollution

2.   Provide answers to the following questions related to *point and area estimates of precipitation, hydrologic cycle processes, surface runoff and ground water flow.* (B.Sc., UoD 2013)

a) Briefly compare and/or contrast the Triangular Area Weighted Mean method with one other method of estimating areal mean precipitation. In your answer provide one key assumption, one advantage and one disadvantage of each method.

b) Identify and briefly discuss three (3) important factors that are necessary to ensure that point or area estimates of precipitation are representative, for a given watershed, regardless of the estimate methods used.

c) Compare and contrast the following hydrologic processes and briefly explain the importance of each component to the hydrologic cycle.

- Evapotranspiration and Transpiration
- Surface runoff and Groundwater flow

3.   Provide answers to the following questions related to *point and areal estimates of precipitation, infiltration simulation, evapotranspiration and the hydrologic equation.* (B.Sc., UoD 2013)

1.   Point precipitation is traditionally measured using various types of rain gages such as the non-retarding cylindrical container type or the recording weighing type, float type and tipping bucket type. Give two (2) reasons why a single point precipitation measurement is typically not representative of the volume of precipitation falling over a given catchment area.

2.   The infiltration rate for excess rain on a small watershed was observed to be 100 mm/h at the beginning of a rainstorm and it decreased exponentially to an equilibrium rate of 10 mm/h after 10 h. A total of I 00 in of water infiltrated during the 10h interval. Determine the value of k in Horton's equation: $f = f_c + (f_o - f_c) * e^{-kt}$

3.   Estimate the amount of evapotranspiration (ET) for the year (mm) from a watershed with a 50,000 km^2 surface area. Consider the drainage area receives 100 mm of rain over the year and the river draining the area has an annual flow rate of 300 m^3/s. Justify any assumptions you make and use the basic equation of hydrology (BEH). Recall that the BEH may be written as:

$P - R - G - E - T = \Delta S$

Where: P = Precipitation, R = Surface runoff, G = Groundwater flow, E = Evaporation, T = Transpiration and ΔS = Change in Storage.

4.   Briefly explain three (3) important components of the Green-Ampt infiltration model used for infiltration modeling.

4.   Provide answers to the following questions (B.Sc., UoD 2013)

    a. International, regional and local organizations, associations and agencies (governmental, NGO's, CBO's) working in aspects of water, wastewater and environment are many such as: UNEP, WHO, CGIAR, FAO, UNESCO, IMO, WFP, WIPO, UNIDO, WMO, UNWTO, UNICEF, GWP, WB, IMF, WWC, ADB, AMCOW, ADB, APWF, GWA, UNHABITAT, USAID, WSSCC, WTO, GEF, EEA, Saudi Environmental Society to name but a few. Since KSA government is a key partner and stakeholder in these entities, show your efforts to maximize use of their resources, knowhow and expertise to address environmental problems and concerns at your locality.

    b) Virtual water is the water embedded in products and used in the whole production chain, and that is traded between regions or exported to other countries. Elaborate on ideas of virtual water transport, trade, dependence and markets in KSA. Suggest methods and ideas to maximize its use.

    c) Integrated water resources management, IWRM, approach has now been accepted internationally as the way forward for efficient, equitable and sustainable development and management of the world's limited water resources and for coping with conflicting demands. Being the responsible environmental engineer and hydrologist at the regulatory and governing agency indicate how you can contribute towards implementation of the concept in KSA. (B.Sc., UoD 2013)

    d) As a district engineer you have been consulted to assist in determining the annual water balance and water budget for KSA based on information regarding volume and distribution of each of: atmospheric water, biological water, rivers, reservoirs, small lakes & farm ponds, marshes, estuaries, soil water and groundwater (fresh and saline). Show how to tackle problems of unavailability and limiting transparency of data and records. What recommendations would you propose to alleviate the situation?

e) Being the environmental engineer in charge of Dammam areas suggest how to improve number and distribution of hydrological stations at your district. What measures would you recommend for monitoring equipment and calibration?

References and bibliography for further reading

1) Abdel-Magid, I. M., & Ibrahim A. A., Hydrology, Sudan University for Science and Technology Publishing House, SUSTPH, Khartoum, Sudan, 2002 (In Arabic).
2) AlHassan, E. E., Adam, O. M. A., Ahmed, M. G. and Abdel-Magid, I.M., Hydrology and hydraulics, submitted for publishing Sudan University for Science and Technology Publishing House, SUSTPH (In Arabic).
3) Bedient, P. B., Huber, W. C. and Vieux, B. E., Hydrology and flood plain analysis, Prentice Hall Inc., Upper Saddle River, NJ, 2008.
4) Bruce, J. P. and Clark, R. H., Introduction to hydrometeorology, Pergamon, Oxford, 1966.
5) Brutsaert, W., Hydrology: An Introduction, Cambridge University Press, NewYork, 2005.
6) Gupta, R. S., Hydrology and hydraulic systems, Waveland Press, Inc., 2001.
7) Hammer, M. J., and MacKichan, K. A., Hydrology and quality of water resources, John Wiley and Sons, NewYork, 1981.
8) Hiscock, K., Hydrogeology: Principles and Practice, Wiley-Blackwell.
9) LaMoreaux P. E., Soliman, M. M, Memon, B. A., LaMoreaux, J. W. and Assaad, F. A., Environmental

hydrology, CRC Press Publishing, Taylor & Francis Group, LLC, Boca Raton, FL, 2009.

10) Linsley, R. K. Kohler, M. A. and Paulhus, J. L. H., Hydrology for Engineers, McGraw-Hill Series in Water Resources and Environmental Engineering, 1982.

11) Linsely, R. K.; Kohler, M. A. and Paulhus, J. L. H., Applied Hydrology, Tata McGraw-Hill Pub. Co., New Delhi, 1983

12) Maidment, D., Handbook of hydrology, McGraw-Hill Professional Publisher; 1993.

13) Raudkivi, A. K., Hydrology – An advanced introduction to hydrological processes and modeling, Pergamon Press, Oxford, 1979.

14) Shaw, E.M., Hydrology in practice, Van Nostrand Reinhold Co., Berkshire, 1983.

15) Subramanya, K. Engineering Hydrology, 3rd Edition, International Edition, McGraw-Hill, London, 2009.

16) Ven Te Chow, Ed., Handbook of Applied Hydrology: A Compendium of Water Resources Technology, McGraw Hill Book Co., New York, 1964.

17) Viessman, W. and Lewis, G. L., Introduction to hydrology, Fifth Edition, Prentice Hall Publisher; 2012.

18) Ward, A. D. and Trimble, S. W., Environmental Hydrology, CRC Press Lewis Publishers, Boca Raton, FL, 2004.

19) Wilson, E. M., Engineering Hydrology, Macmillan Education, London, 1990.

20) Wisler, C. O. and Brater, E. F., Hydrology, John Wiley and Sons, New York, 1959.

21) WMO, Guide to Hydrological Practices, 5th ed., WMO-No. 168. Chapters 20, 21, 1994.

22) WMO, Terakawa, A., Hydrological Data Management: Present State and Trends, WMO-No. 964. 2003, Available through http://www.wmo.int/.

23) WMO, Guide to Meteorological Instruments and Methods of Observation, 2008, Available through http://www.wmo.int/.

24) Faris, G.F. and Abdel-Magid, I.M., Fundamentals of engineering hydrology, under preparation.

Useful Formulae

1) $e = P - P'$

2) $e_w - e = \square \, (t - t_w)$

3)
$$P_x = \frac{\dfrac{N_x P_a}{N_a} \cdot \dfrac{N_x P_b}{N_b} \cdot \dfrac{N_x P_c}{N_c}}{3}$$

$$P_{mean} = \frac{A_1}{A} P_1 + \frac{A_2}{A} P_2 + \ldots\ldots + \frac{A_n}{A} P_n = \left(\sum_{i=1}^{n} A_i P_i \right) \div \left(\sum_{i=1}^{n} A_i \right)$$

4)
$$P_{av} = \sum_{i=1}^{n} \frac{P_i}{n}$$

5)
$$P_{mean} = \frac{\displaystyle\sum_{i=1}^{n} A_i P_i}{\displaystyle\sum_{i=1}^{n} A_i}$$

6) $i = \dfrac{a}{t+b}$

7) $i = \dfrac{c}{t^n}$

8) $i = \dfrac{60 P}{t}$

9) $P = \left[\dfrac{1.214 \times 10^5}{600} \times N_t \right]^{0.282} - 2.54$

10) $\dfrac{\overline{P}}{P} = 1 - \dfrac{0.3\sqrt{A}}{t^\upsilon}$

11) $i = \left[\dfrac{60}{t} \right] \times \left[(202.3 \, N_t)^{0.282} - 2.54 \right]$

12) $EV = b(e_s - e)$

13) $EV = f(u). (e_s - e)$

14)
$$E_T = \frac{\Delta H + \gamma E_a}{\Delta + \gamma}$$

15)
$$E_T = E_1\left(t, \frac{n}{D}\right) + E_2\left(t, R_A, \frac{n}{D}\right) + E_3\left(t, \frac{n}{D}, h\right) + E_4\left(t, u_2, h\right)$$

16) $f = f_c + (f_o - f_c)e^{-kt}$

17) $F = \int\limits_0^t fdt = f_c t + \left[\frac{(f_0 - f_c)}{k_f}\right]\left(1 - e^{k_f t}\right)$

18) $F_c = \frac{(f_0 - f_c)}{k_f}$

19) $I_t = I_o * k^t$

20) $v = k*i$

21) $\bar{v} = \frac{v}{n_e}$

22) $v = -k\frac{d\varphi}{dx}$

23) $q = -kH.\frac{d\varphi}{dx}$

24) $\varphi = -\frac{v.x}{k}$

25) $q = -kh\frac{dh}{dx}$

26) $q = \frac{k\left(h_0^2 - h^2\right)}{2x}$

27) $h^2 = -\frac{N x^2}{k} + ax + b$

28) $s_1 - s_2 = \frac{Q_o}{2\pi kH}\ln\left[\frac{r_2}{r_1}\right]$

29) $k = \frac{Q}{2\pi H\left[S_1 - S_2\right]}\ln\left[\frac{r_2}{r_1}\right]$

30) $T = kH = \dfrac{Q}{2\pi[S_1 - S_2]} \ln\left[\dfrac{r_2}{r_1}\right]$

31) $S = \dfrac{Q_o}{2\pi kH} \ln\dfrac{R_o}{r} = \dfrac{Q_o}{2\pi T} \ln\dfrac{R_o}{r}$

32) $Q_o = \dfrac{\pi k[H^2 - h_o^2]}{\ln\dfrac{R}{r_o}}$

33) $H^2 - h^2 = \dfrac{Q_o}{\pi k} \ln\dfrac{R}{r} - \dfrac{N}{2k}[R^2 - r^2]$

34) $Q = \pi R^2 N$

35) $Q = 27.78\ C\,I\,A$

36) $Q = 2.78\ C\,I\,A$

37) $i = \dfrac{cT^m}{(t+d)^n}$

38) $\dfrac{Q_a}{Q} = \sqrt{\left[1 + \dfrac{1}{US}\dfrac{dh}{dt}\right]}$

39) $U = 1.3\dfrac{Q_a}{A}$

40) $\dfrac{Q_a}{Q} = \sqrt{\left[1 + \dfrac{A}{1.3\,Q_a\,S}\dfrac{dh}{dt}\right]}$

41) $Q = k\,(h - a)^{\,x}$

42) $Q = AC\sqrt{r_H . S}$

43) $I - D = \dfrac{\Delta S}{\Delta t}$

44) $N = b.A^{0.2}$

45) $0.5\,(I_1 + I_2)\Delta t - 0.5\,(D_1 + D_2)\Delta t = S_2 - S_1$

46) $S = K[XI + (1 - X)D]$ 47)

Appendices

NOMOGRAM FOR DETERMINING EVAPORATION E_0 FROM A FREE WATER SURFACE ACCORDING TO THE FORMULA OF PENMAN

$$E_0 = \frac{\Delta H + \gamma E a}{\Delta + \gamma} = E_1\left(t, \frac{n}{D}\right) + E_2\left(t, R_A, \frac{n}{D}\right) + E_3\left(t, \frac{n}{D}, h\right) + E_4(t, u_2, h)$$

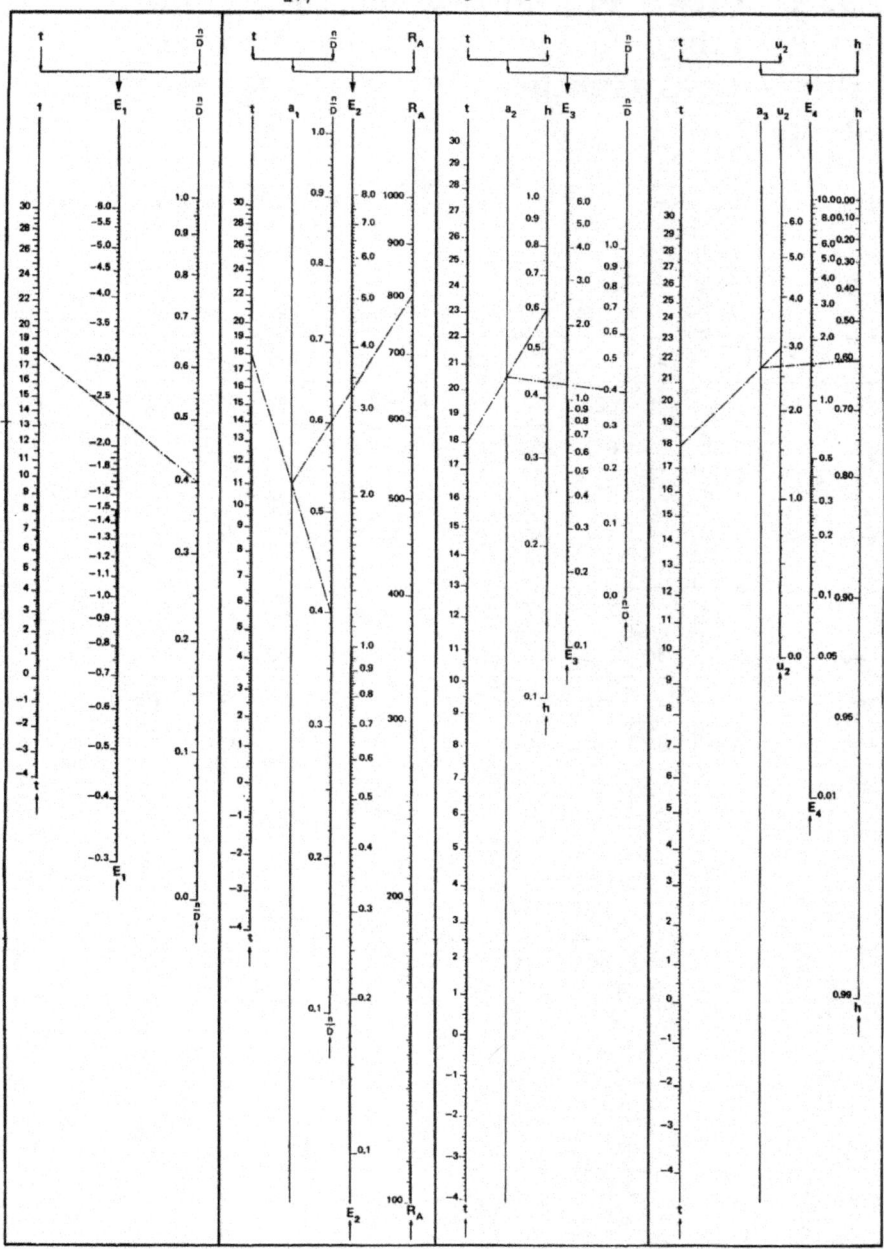

Appendix (B) Agot's values of short-wave radiation flux R_A at the outer limit of the atmosphere in g cal/cm^2/day as a function of the month of the year & latitude.

Latitude (Degrees)	Jan	Feb	Mar	Apr	May	Jun	Jul	Aug	Sep	Oct	Nov	Dec	Yr
N 90	0	0	55	518	903	1077	944	605	136	0	0	0	3540
80	0	3	143	518	875	1060	930	600	219	17	0	0	3660
60	86	234	424	687	866	983	892	714	494	258	113	55	4850
40	38	538	663	847	930	1001	941	843	719	528	397	318	6750
20	631	795	821	914	912	947	912	887	856	740	666	599	8070
Equator	844	963	878	876	803	803	792	820	891	866	873	829	8540
20	970	1020	832	737	608	580	588	680	820	892	986	978	8070
40	998	963	686	515	358	308	333	453	648	817	994	1033	670
60	947	802	459	240	95	50	77	187	403	648	920	1013	4850
80	981	649	181	9	0	0	0	0	113	459	917	1094	3660
S 90	995	656	92	0	0	0	0	0	30	447	932	1110	3540

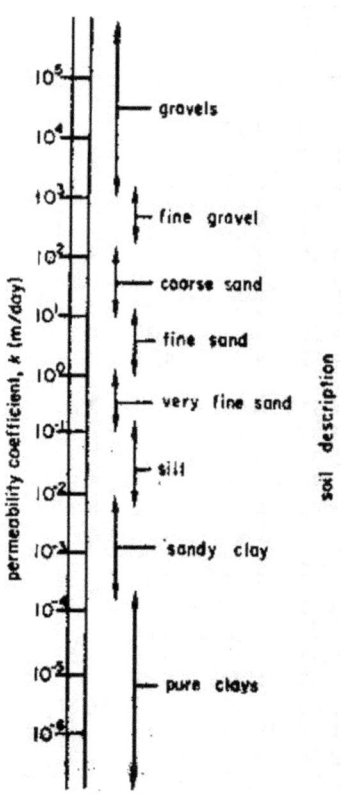

Appendix (C) Soil descriptions with permeability coefficient.

Soil description	Permeability coefficient, m/day
Pure clays	10^{-4} to 10^{-7}
Sandy clay	10^{-2} to 10^{-4}
Silt	10^{-1} to 10^{-2}
Very fine sand	10^{-1} to 1
Coarse sand	1 to 10^{2}
Fine gravel	10^{2} to 10^{3}
Gravels	More than 10^{3}

t (°C)	e_s									
	0.0	0.1	0.2	0.3	0.4	0.5	0.6	0.7	0.8	0.9
−10	2.15									
−9	2.32	2.30	2.29	2.27	2.26	2.24	2.22	2.21	2.19	2.17
−8	2.51	2.49	2.47	2.45	2.43	2.41	2.40	2.38	2.36	2.34
−7	2.71	2.69	2.67	2.65	2.63	2.61	2.59	2.57	2.55	2.53
−6	2.93	2.91	2.89	2.86	2.84	2.82	2.80	2.77	2.75	2.73
−5	3.16	3.14	3.11	3.09	3.06	3.04	3.01	2.99	2.97	2.95
−4	3.41	3.39	3.37	3.34	3.32	3.29	3.27	3.24	3.22	3.18
−3	3.67	3.64	3.62	3.59	3.57	3.54	3.52	3.49	3.46	3.44
−2	3.97	3.94	3.91	3.88	3.85	3.82	3.79	3.76	3.73	3.70
−1	4.26	4.23	4.20	4.17	4.14	4.11	4.08	4.05	4.03	4.00
−0	4.58	4.55	4.52	4.49	4.46	4.43	4.40	4.36	4.33	4.29
0	4.58	4.62	4.65	4.69	4.71	4.75	4.78	4.82	4.86	4.89
1	4.92	4.96	5.00	5.03	5.07	5.11	5.14	5.18	5.21	5.25
2	5.29	5.33	5.37	5.40	5.44	5.48	5.53	5.57	5.60	5.64
3	5.68	5.72	5.76	5.80	5.84	5.89	5.93	5.97	6.01	6.06
4	6.10	6.14	6.18	6.23	6.27	6.31	6.36	6.40	6.45	6.49
5	6.54	6.58	6.54	6.68	6.72	6.77	6.82	6.86	6.91	6.96
6	7.01	7.06	7.11	7.16	7.20	7.25	7.31	7.36	7.41	7.46
7	7.51	7.56	7.61	7.67	7.72	7.77	7.82	7.88	7.93	7.98
8	8.04	8.10	8.15	8.21	8.26	8.32	8.37	8.43	8.48	8.54
9	8.61	8.67	8.73	8.78	8.84	8.90	8.96	9.02	9.08	9.14
10	9.20	9.26	9.33	9.39	9.46	9.52	9.58	9.65	9.71	9.77
11	9.84	9.90	9.97	10.03	10.10	10.17	10.24	10.31	10.38	10.45
12	10.52	10.58	10.66	10.72	10.79	10.86	10.93	11.00	11.08	11.15
13	11.23	11.30	11.38	11.75	11.53	11.60	11.68	11.76	11.83	11.91
14	11.98	12.06	12.14	12.22	12.96	12.38	12.46	12.54	12.62	12.70
15	12.78	12.86	12.95	13.03	13.11	13.20	13.28	13.37	13.45	13.54
16	13.63	13.71	13.80	13.90	13.99	14.08	14.17	14.26	14.35	14.44
17	14.53	14.62	14.71	14.80	14.90	14.99	15.09	15.17	15.27	15.38
18	15.46	15.56	15.66	15.76	15.96	15.96	16.06	16.16	16.26	16.36
19	16.46	16.57	16.68	16.79	16.90	17.00	17.10	17.21	17.32	17.43
20	17.53	17.64	17.75	17.86	17.97	18.08	18.20	18.31	18.43	18.54
21	18.65	18.77	18.88	19.00	19.11	19.23	19.35	19.46	19.58	19.70
22	19.82	19.94	20.06	20.19	20.31	20.43	20.58	20.69	20.80	20.93
23	21.05	21.19	21.32	21.45	21.58	21.71	21.84	21.97	22.10	22.23
24	22.27	22.50	22.63	22.76	22.91	23.05	23.19	23.31	23.45	23.60
25	23.75	23.90	24.03	24.20	24.35	24.49	24.64	24.79	24.94	25.08
26	25.31	25.45	25.60	25.74	25.89	26.03	26.18	26.32	26.46	26.60
27	26.74	26.90	27.05	27.21	27.37	27.53	27.69	27.85	28.00	28.16
28	28.32	28.49	28.66	28.83	29.00	29.17	29.34	29.51	29.68	29.85
29	30.03	30.20	30.38	30.56	30.74	30.92	31.10	31.28	31.46	31.64
30	31.82	32.00	32.19	32.38	32.57	32.76	32.95	33.14	33.33	33.52

Appendix (D) Saturation vapour pressure as a function of temperature t (Negative values of t refer to conditions over ice; 1 mm Hg = 1.33 mbar)

Appendix (E)
Some Physical Properties of Water at Various Temperatures.

Temp, °C	Density, kg/m^3	Dynamic viscosity μx10^3 Ns/m^2	Kinematic viscosity vx10^6 m^2/s
0	999.8	1.793	1.792
5	1000	1.519	1.519
10	999.7	1.308	1.308
15	999.1	1.140	1.141
20	998.2	1.005	1.007
25	997.1	0.894	0.897
30	995.7	0.801	0.804
35	994.1	0.723	0.727
40	992.2	0.656	0.661
45	990.2	0.599	0.605
50	988.1	0.549	0.556
55	985.7	0.506	0.513
60	983.2	0.469	0.477
65	980.6	0.436	0.444
70	977.8	0.406	0.415
75	974.9	0.380	0.390
80	971.8	0.357	0.367
85	968.6	0.336	0.347
90	965.3	0.317	0.328
95	961.9	0.299	0.311
100	958.4	0.284	0.296

Appendix (F)
Conversion Table

Multiply	By	to obtain
Area		
acre	43560	ft^2
acre	4047	m^2
cm^2	0.155	in^2
ft^2	0.0929	m^2
hecatre, ha	2.471	acre
in^2	6.452	cm^2
km^2	0.3861	$mile^2$
m^2	10.76	ft^2
mm^2	0.00155	in^2
Concentration		
mg/l	8.345	lb/million USA gal
ppm	1	mg/L
Density		
g/cm^3	1000	kg/m^3
g/cm^3	1	kg/L
g/cm^3	62.43	lb/ft^3
g/cm^3	10.022	lb/gal (Br.)
g/cm^3	8.345	lb/gal (USA)
kg/m^3	0.001	g/cm^3
kg/m^3	0.001	kg/L
kg/m^3	0.6242	lb/ft^3
Flowrate		
ft^3/s	448.8	gal/min
ft^3/s	28.32	L/s
ft^3/s	0.6462	M gal/d
M gal/d	1.547	ft^3/s
gal/min	0.00223	ft^3/s
gal/min	0.0631	L/s
L/s	15.85	gal/min
m^3/hr	4.4	gal/min
m^3/s	35.31	ft^3/s

Length		
ft	30.48	cm
in	2.54	cm
km	0.06214	mile
km	3280.8	ft
m	39.37	in
m	3.281	ft
m	1.094	yard
mile	5280	ft
mile	1.6093	km
mm	0.03937	in
Mass		
g	$10^{-3}*2.205$	lb
lb	16	ounce
kg	2.205	lb
tonne, t	1.102	ton (2000 lb)
Power		
Btu	252	cal
Btu	778	ft-lb
Btu	$10^{-4}*3.93$	HP-hr
Btu	$10^{-4}*2.93$	kW-hr
HP	0.7457	kW
Pressure		
atm	760	mm Hg
atm	29.92	in Hg
atm	33.93	ft water
atm	10.33	m water
atm	$10^{4}*1.033$	kg/m^2
atm	$10^{5}*1.013$	N/m^2
in water	1.8665	mm Hg
in Hg	0.49116	lb/in^2
in Hg	25.4	mm Hg
k Pa	0.145	lb/in^2, psi
mm Hg	0.01934	lb/in^2
mm Hg	13.595	kg/m^2
lb/in^2	6895	N/m^2
lb/in^2	0.0703	kg/cm^2

Temperature		
Fahrenheit, F	9/(F – 32)*5	Centigrade, C
Centigrade, C	32 + (9C/5)	F
C	C + 273.16	Kelvin, K
Rankine	F + 459.69	F
Velocity		
ft/s	30.48	cm/s
ft/s	1.097	km/hr
ft/min	0.508	cm/s
cm/s	0.03281	ft/s
cm/s	0.6	m/min
m/s	3.281	ft/s
m/s	196.8	ft/min
Viscosity		
centipoise	0.01	g/cm.s
centistoke	0.01	cm^2/s
Volume		
ft^3	0.02832	m^3
ft^3	6.229	gal (Br.)
ft^3	7.481	gal (USA)
ft^3	28.316	L
gal	0.1337	ft^3
gal (USA)	0.833	gal (Br.)
gal	3.785	L
L	0.001	m^3
L	0.03532	ft^3
L	0.22	gal (Br.)
L	0.2642	gal (USA)
m^3	35.314	ft^3
m^3	1000	L

Appendix (G)
Screenshots of the example programs

Program 2.1 – Form1.vb (Design):

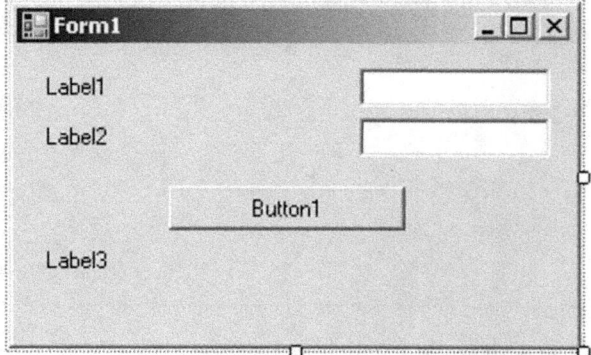

Program 2.2 – Form1.vb (Design):

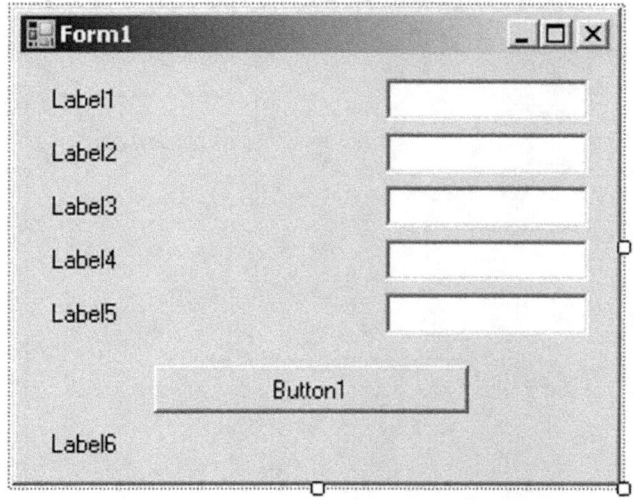

Program 2.3 – Form1.vb (Design):

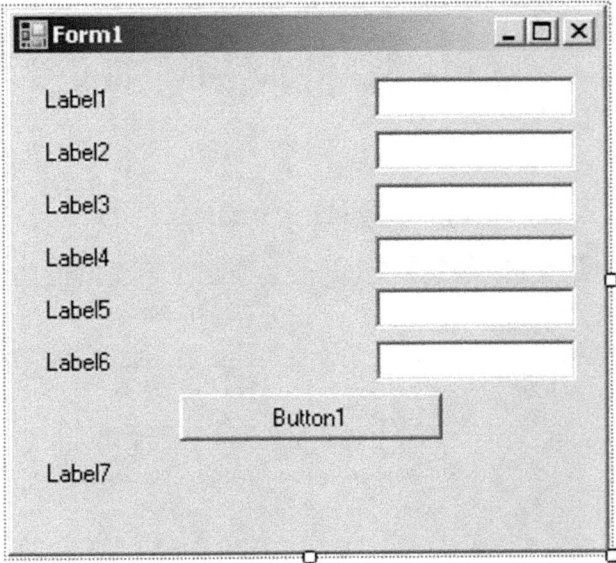

Program 2.4 – Form1.vb (Design):

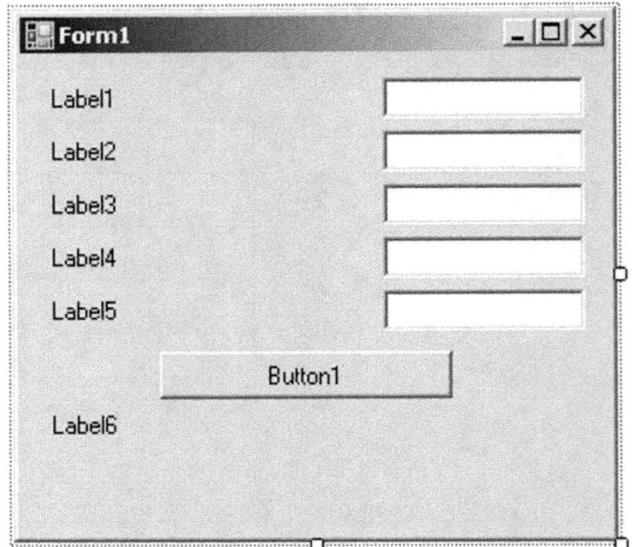

Program 3.1 – Form1.vb (Design):

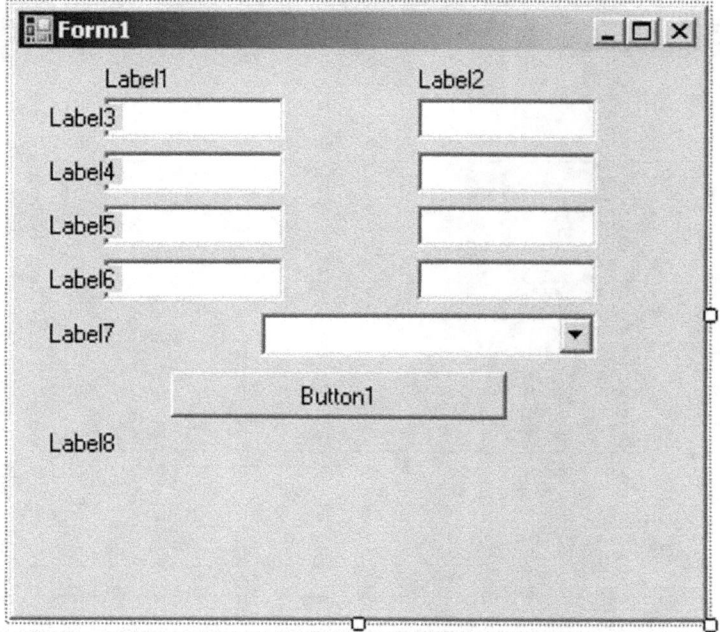

Program 3.4 – Form1.vb (Design):

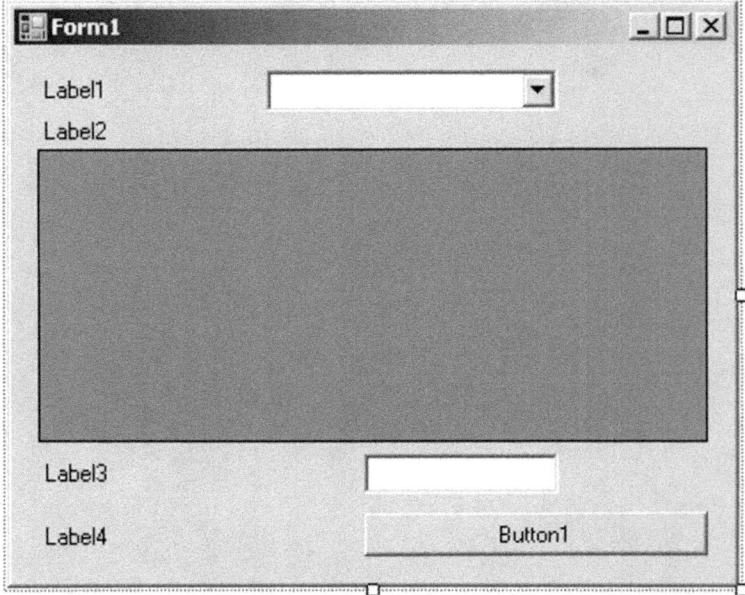

Program 3.6 – Form1.vb (Design):

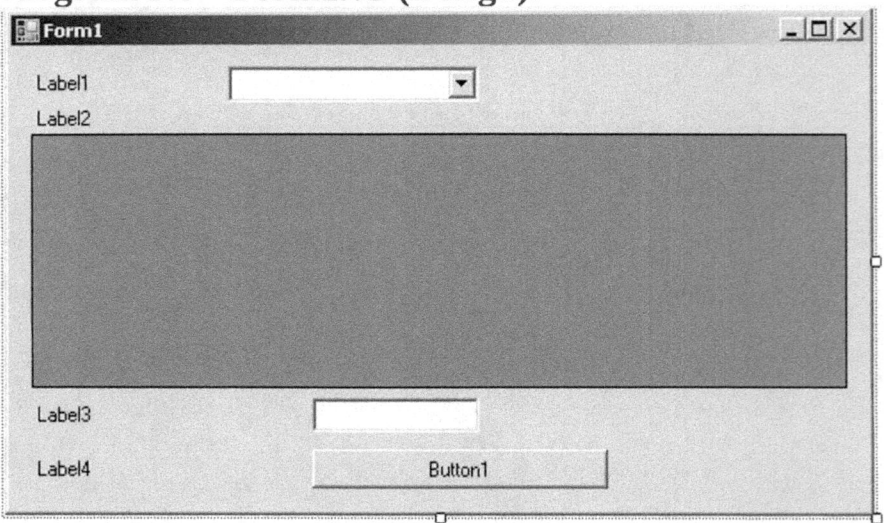

Program 4.2 – Form1.vb (Design):

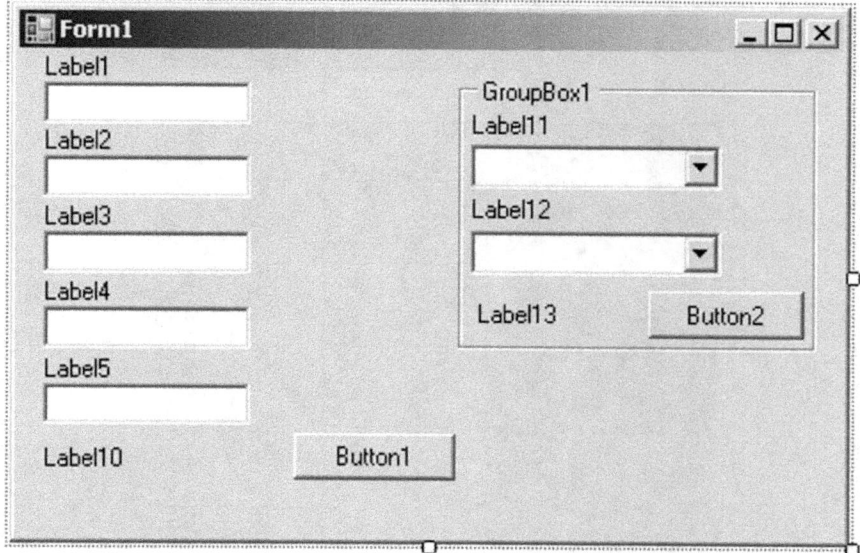

Program 5.1 – Form1.vb (Design):

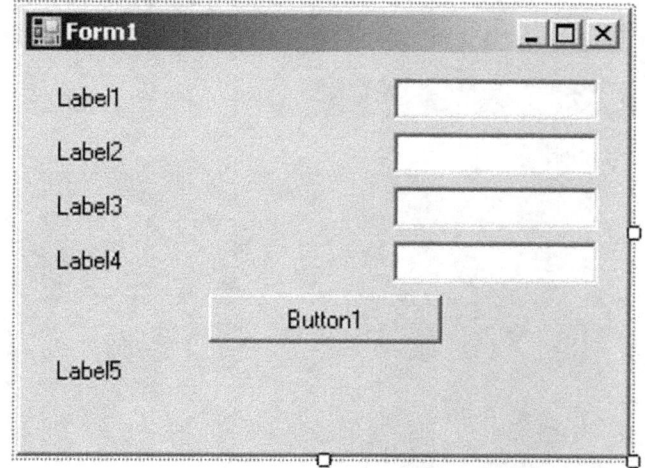

Program 5.2 – Form1.vb (Design):

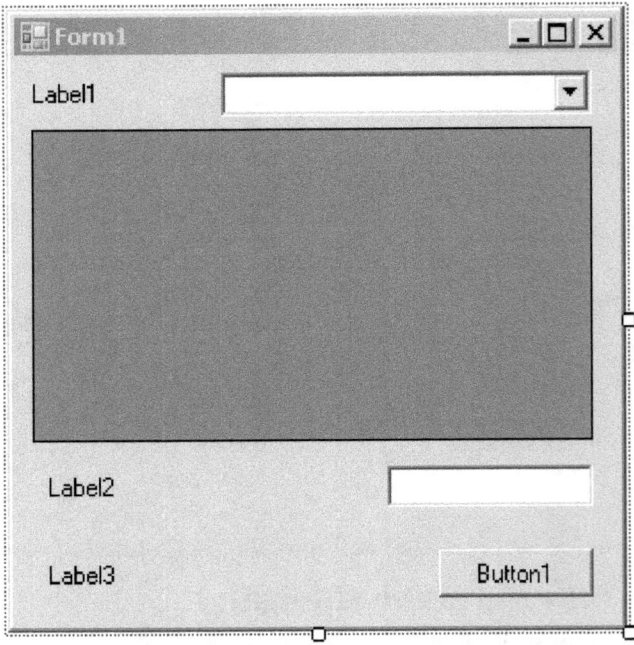

Program 5.4 – Form1.vb (Design):

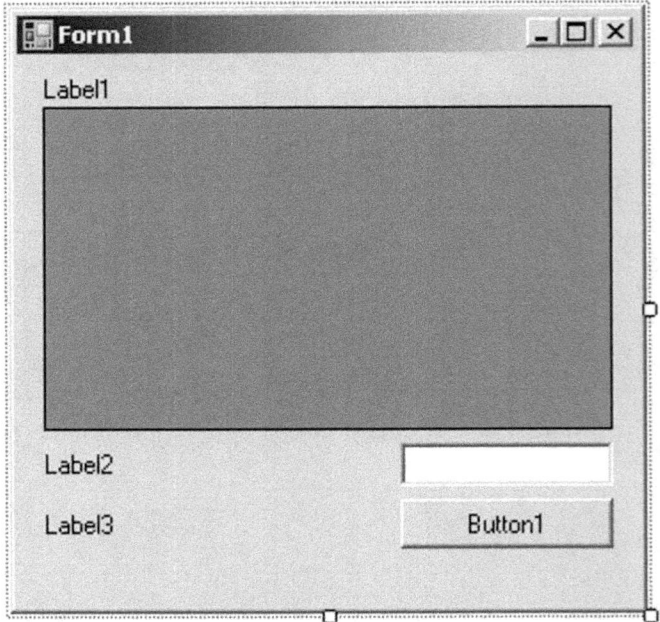

Program 5.6 – Form1.vb (Design):

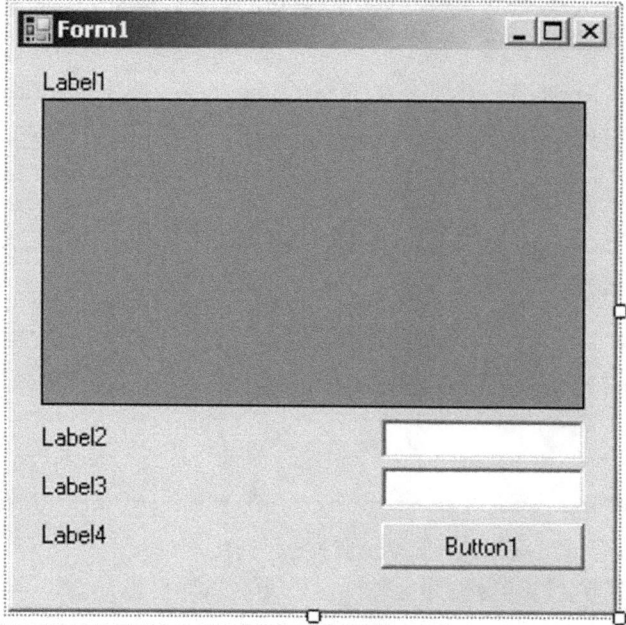

Program 6.1 – Form1.vb (Design):

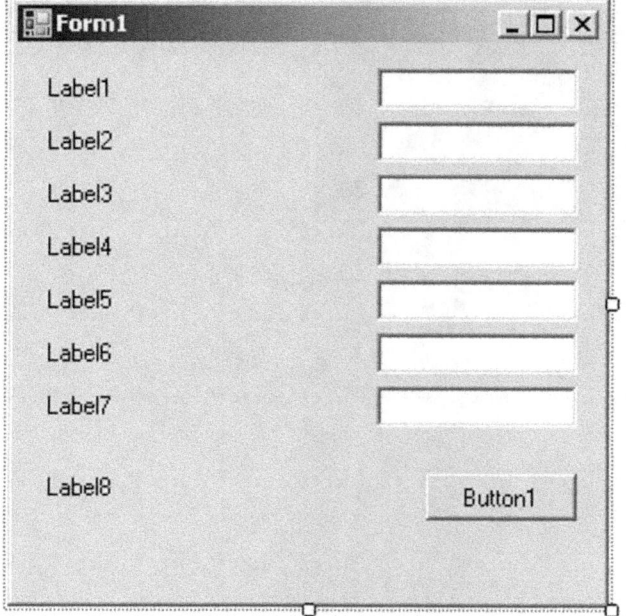

Program 6.2 – Form1.vb (Design):

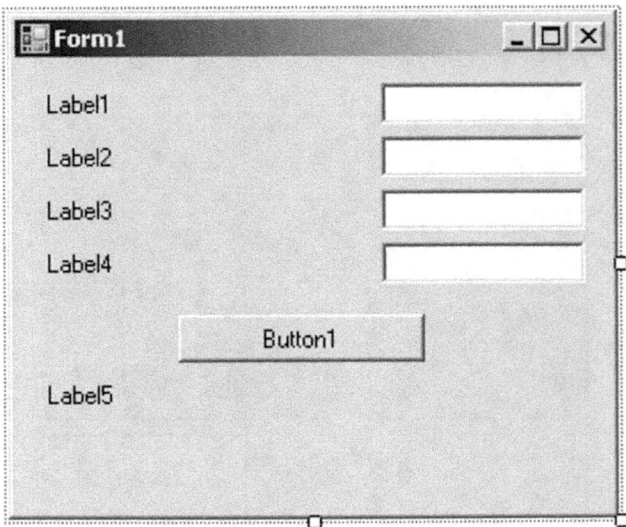

Program 7.1 – Form1.vb (Design):

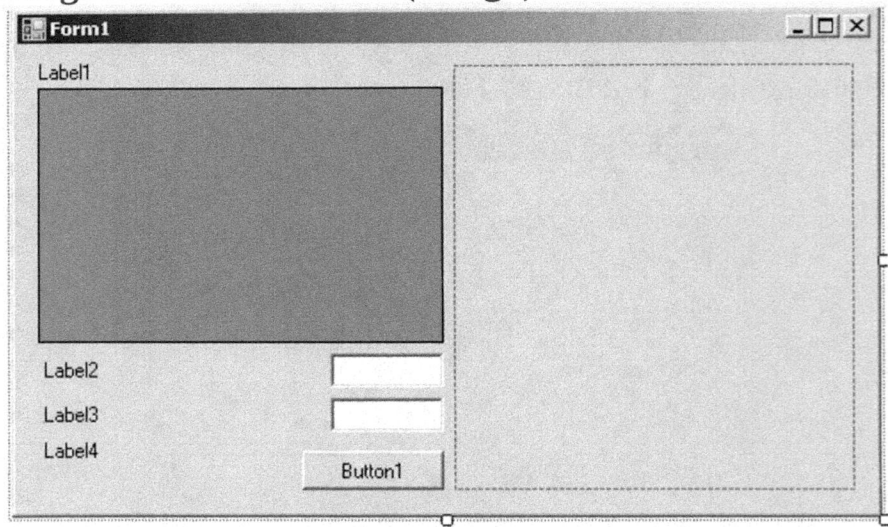

Program 8.1 – Form1.vb (Design):

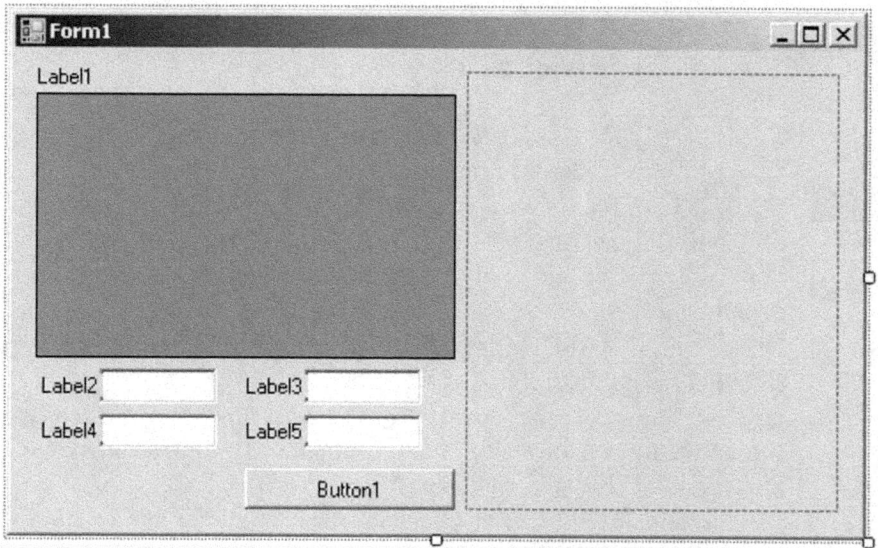

About the authors

Assoc. Prof. Dr. **Faris Gorashi Faris,** B.Sc. M.Sc., Ph.D., MASCE, MIWA, MEWB.

1) Associate Professor at Infrastructure University Kuala Lumpur.

2) Head of Post-Graduate studies in the Department of Civil Engineering at IUKL. Worked as: Head of Water and Wastewater Bachelor of Technology program at the same department.

3) Director of IKRAM Centre of Advanced Material and Technology (ICAMT) Kumpulan IKRAM Sdn Bhd.

4) Deputy Dean of the School of Engineering and Technology Infrastructure at University Kuala Lumpur, (IUKL), Jalan Ikram – Uniten, 43000 Kajang , Selangor MA.

5) Vice principal IIUM Lower Education Group

6) Has over 15 years of experience as a water resources engineer, and hydrologist. His expertise includes water network design and simulation, water quality modeling, waste water reclamation and reuse, and environmental impact assessment and general management. He has worked with multi-national companies and organizations, holding various technical, research, general management and consultancy posts.

7) Holds a doctoral degree in Built Environment from the Faculty of Architecture and Environmental Design of the Islamic University Malaysia (2009), a Master of Engineering in Hydrology (2002) and a BSc Honors in Civil Engineering from Omdurman Islamic University.

8) Has published a number of journal papers and developed the LA-WQI model. He has also held several lecturer and administrative posts at a number of institutions of education

Prof. Dr. Eng. **Isam Mohammed Abdel-Magid Ahmed**, B.Sc. PDH, DDSE, Ph.D, FSES, CSEC (Consultant engineer EC/ER/CE/146), MIWEM, MIWRA, MGWPEASC.

1) B.Sc., Honors (first class), University of Khartoum, Faculty of Eng., Civil Eng. Dept., (Sudan) May, 1977. Diploma Hydrology, Padova University (Italy), July, 1978. Diploma in Sanitary Eng., Delft University of Technology (The Netherlands), equivalent to M.Sc., September, 1979. Ph.D., Public Health Eng., Strathclyde University (Great Britain), June, 1982

2) Professor of Water Resources and Environmental Engineering.

3) Worked at: General Corporation for Irrigation and Drainage (Sudan), University of Khartoum (Sudan), University of United Arab Emirates (United Arab Emirates), Sultan Qaboos University (Oman), Omdurman Islamic University, Sudan University for Science and Technology, Juba University, Industrial Research and Consultancy Center, Sudan Academy of Sciences of the Ministry of Science and Technology (Sudan), King Faisal University (Saudi Arabia) and University of Dammam (Saudi Arabia).

4) Supervised many projects and dissertations leading to different degrees: Diploma, B.Sc., and Higher diploma, M.Sc., and Ph.D. Selected external examiner to different institutions. Actively acted and participated in different committees, boards, societies, institutions, and organizations.

5) Reviewed many articles, papers and research studies for different authorities. Actively participated in or attended many local, regional and international conferences, congresses, seminars and workshops.

6) Obtained the Sudan Engineering Society Prize for the Best Project in Civil Engineering, 7th June 1977. The Ministry of Irrigation Prize for Second Best Performance in 4th year Civil Engineering, 7th June 1977 and the Honourly Scarf for Enrichment of Knowledge, Khartoum University Press, February the 25th 1986, Abdallah Eltaib prize for the best published book, 2000.

7) Authored or co-authored over 100 publications, many text and reference books, many technical reports and lecture notes in areas of: water supply; wastewater disposal, reuse and reclamation; solid waste disposal; water resources and management. Edited or co-edited many local and international conference proceedings and college bulletins.

 **Dr. Mohammed Isam Mohammed Abdel-Magid**

1) Dr. Mohammed Isam Mohammed Abdel-Magid (MBBS, BLS, ALS, MRCP-UK Part I&II Written) is a graduate of the College of Medicine, University of Khartoum, Sudan, 2008. He completed basic training with the Ministry of Health, Sudan, then worked as a physician in the department of Internal Medicine, Ribat University hospital, Sudan, and the Ministry of Health, Kingdom of Saudi Arabia.

2) He completed his higher training with the membership of the Royal Colleges of Physicians of the United Kingdom (MRCP-UK) of which he completed its three parts.

3) He tutored in problem-based learning teaching sessions in the department of Internal Medicine, Sudan International University, Sudan.

4) He is a registered practicing physician with the Sudan Medical Council, the Health Authority of Abu-Dhabi (HAAD), and the Saudi Commission of Health Specialties (SCHS). He is a full member of the Society of Acute Medicine of UK (SAM), the European Society for Emergency Medicine (EuSEM), and the European Respiratory Society (ERS).

5) He is a peer reviewer with the Science Journal of Medicine & Clinical Trial and the Pan-African Journal of Medical Sciences.

www.ingramcontent.com/pod-product-compliance
Lightning Source LLC
Chambersburg PA
CBHW051904170526
45168CB00001B/234